Nursing care

Nursing care

A HOMEOSTATIC CASEBOOK

Edited by

John Clancy BSc(Hons), PGCEA
Lecturer in Physiology Applied to Health,
University of East Anglia, Norwich, UK

Andrew McVicar BSc(Hons), PhD
Senior Lecturer in Physiology Applied to Health,
Anglia Polytechnic University, Chelmsford, UK

ARNOLD

A member of the Hodder Headline Group
LONDON • NEW YORK • SYDNEY • AUCKLAND

First published in Great Britain in 1998 by
Arnold, a member of the Hodder Headline Group,
338 Euston Road, London NW1 3BH
http://www.arnoldpublishers.com

Whilst the advice and information in this book is believed to be true and
accurate at the date of going to press, neither the authors nor the publisher
can accept any legal responsibility or liability for any errors or omissions
that may be made.

British Library Cataloguing in Publication Data
A catalogue record for this book is available from the British Library

Library of Congress Cataloging-in-Publication Data
A catalog record for this book is available from the Library of Congress

ISBN 0 340 69255 3 ✓

Publisher: Fiona Goodgame
Production Editor: Wendy Rooke
Production Controller: Sarah Kett
Cover Designer: Terry Griffiths

Typeset in 10/12pt Palatino by
Scribe Design, Gillingham, Kent
Printed and bound in Great Britain by
The Bath Press, Bath

Contents

Contributors

Theresa Atherton, BSc(Hons), RGN, RSCN, DN(London), CertEd.,
Lecturer in Children's Health, School of Health (Nursing and Midwifery), University of East Anglia, Norwich

Roy Bishop BA(Hons), RMN, CertEd., RNT,
Lecturer in Human Sciences School of Health (Nursing and Midwifery), University of East Anglia, Norwich

Stevie Boyd, RGN, RSCN,
Special Care Baby Unit, Norfolk and Norwich Hospital, Norwich

John Clancy, BSc(Hons), PGCEA,
Lecturer in Physiology Applied to Health, School of Health (Nursing and Midwifery), University of East Anglia, Norwich

Gibson D'Cruz, RGN, RNT, BEd(Hons), BA, MSc.,
Lecturer in Adult Care, School of Health (Nursing and Midwifery), University of East Anglia, Norwich

Elaine Domek, BA, PGCE, RGN, RSCN, ONC,
Lecturer in Children's Health, School of Health (Nursing and Midwifery), University of East Anglia, Norwich

Louise Fuller, BA, RGN, OND,
Practice Nurse, Mundesley Medical Centre, Mundesley, Norfolk

Carolyn Galpin, BSc(Hons), RGN, RCNT, RNT, DipEdN,
Lecturer in Nursing, School of Health (Nursing and Midwifery), University of East Anglia, Norwich

Sally Hardy, MSc, BA(Hons), DPNS, RMN, RGN,
Lecturer in Human Sciences, School of Health (Nursing and Midwifery), University of East Anglia, Norwich

Julia Hubbard, BSc(Hons), RGN, PGDHE, RNT
Lecturer in Coronary Care, School of Health (Nursing and Midwifery), University of East Anglia, Norwich

Gabriel Ip, RGN, RNMH, RNT, DipNS, BSc., Cert Ed
Lecturer in Learning Disabilities, School of Health (Nursing and Midwifery), University of East Anglia

Andrew McVicar, BSc(Hons), PhD,
Senior Lecturer in Physiology Applied to Health, School of Health Care Practice, Anglia Polytechnic University, Chelmsford

Janice Mooney, RGN, DMS,
Rheumatology Nurse Practitioner, Norfolk and Norwich Hospital, Norwich

Christine Nightingale, RNMH, DN, CertEd.,
Lecturer in Learning Disabilities, School of Health (Nursing and Midwifery), University of East Anglia, Norwich

Maggie Quinn, BA(Hons), RGN, PGCE,
Lecturer, School of Health (Nursing and Midwifery), University of East Anglia, Norwich

Joan Schulz, PGD Applications in Psychology, RMN, RGN, CPNCert, Dip. Nursing, Diploma Counselling, CertEd(FE), RCNT,
Lecturer in Mental Health, School of Health (Nursing and Midwifery), University of East Anglia, Norwich

Derek Shirtliffe, BSc(Hons), Dip.N., RNMH, CertEd., Cert Child and Adolescent Psychiatry,
Lecturer, School of Health (Nursing and Midwifery), University of East Anglia, Norwich

Sue Sides, BSc(Hons), RGN, DN, CertEd,
Lecturer in Adult Care, School of Health (Nursing and Midwifery), University of East Anglia, Norwich

Helen Thomas, BSc (Hons), RGN, PGCEA,
Lecturer in Critical Care Nursing, School of Health (Nursing and Midwifery), University of East Anglia, Norwich

Judith Tyler, BA(Hons), RGN, RM, RHV, CertEd, Mphil, PhD,
Senior Lecturer and Team Leader in Primary Care and Midwifery, School of Health (Nursing and Midwifery), University of East Anglia, Norwich

Acknowledgements

Many people, knowingly and unknowingly, have contributed to the production of this book: nursing lecturers, nursing staff, students, family, and friends. We are grateful to them all.

In particular our thanks go to:

The contributors. All of them are nurses experienced in teaching nurses. Their expertise has made the editing of this book relatively easy. Congratulations everyone. Most co-authors are lecturers from the University of East Anglia; three are in fact former students and a special thanks and congratulations goes to them.

Penny, Clare, Lisa and Rachel, whose continued support and lack of complaint (well almost) of our working long, and often unsociable hours, has meant that we completed editing this book in record time.

Our parents and friends who have seen less of us as a result of taking on an increased workload. Who knows: they may be grateful of the rest!

Our colleagues, David Knock and Colleen Bale, for their help with the presentation of various forms of the homeostatic graph – honest they are necessary!

The University of East Anglia and Anglia Polytechnic University for help with resources.

Finally, Fiona Goodgame, Carol Goodall and Clare Parker at Arnold for their advice and support.

John Clancy
Andrew McVicar
1998

List of figures

About this book

First of all, why 'homeostasis'? In its basic principles, homeostasis is a simple model of how the human body is *controlled*. It is a concept that can be applied to all body functions. In its detail, however, the precise mechanisms involved in regulating those functions are complex. Indeed, research continues to raise this complexity still higher. Nevertheless, an appreciation of homeostatic principles provides a framework on which to base learning, and a means to understanding how disorders arise and how the signs and symptoms of a disorder can be so diverse.

Second, why a *casebook*? Applying theory to practice reinforces understanding: 'homeostasis' remains a concept unless the practitioner can relate it to clinical practice. Cases which illustrate the homeostatic disturbances evident in clinical disorders help to emphasize the need for control processes, and also show how clinical intervention is targeted at restoring homeostasis. In this sense, nursing, medical and other health-care interventions act as extrinsic homeostatic mechanisms that are directed at facilitating normal function for the patient/client. This casebook, therefore, places homeostasis as the central concept in health and health care, and *the reader should consider the principles outlined in the Introductory section before proceeding with the book* as these determine the homeostatic approach used in the text.

The term 'homeostasis' has its origins in physiology and is often seen as being of particular relevance to physical wellbeing. This is erroneous: the principle relates just as much to psychological equilibrium and to the psychophysiological processes involved. To emphasize this, *Nursing care: a homeostatic casebook* has drawn on the expertise of nurse teachers and experienced nurse practitioners with specialist interests to produce a range of examples of case studies from within adult care, child care, mental health and learning difficulties. The examples given are not intended to be exhaustive but serve to illustrate the theme. A simple graph is used throughout to reinforce the principles involved; it first appears in the Introductory section and its application will become increasingly apparent as the reader progresses through the book.

It is not the intention of this book to explain in detail any underlying physiology in relation to these examples. Where appropriate, cross-referencing is provided to direct the reader to relevant sections or passages within the editors' published textbook *Physiology and anatomy: a homeostatic approach* (1995).That book also follows the homeostasis theme, applies similar graphs to physiological processes, and supports the approach taken in this casebook.

One final note. Although this book may convey an impression that homeostatic failure promotes clinical disorder, this is not always the case. Should your concentration lapse while reading this book, or your head ache, or perhaps parts of your anatomy ache because you have been sitting too long, remember that these too are signs that local homeostasis has been compromised in these tissues and it is time for a rest to allow recovery! Hopefully, the book will not have such an effect and you find it to be a useful addition to support your studies.

Happy reading!

John Clancy
Andrew McVicar
Editors

Introductory section

Homeostasis: the key concept to nursing care

John Clancy and Andrew McVicar

Introduction

'Homeostasis' refers to the self-regulating physiological processes necessary to maintain the normal state of the body's internal environment. Collectively, physiological function and the maintenance of homeostasis enable the body to attain the basic needs necessary for 'healthy' life.

The human body consists of trillions of microscopic cells and each cell is regarded as a 'basic unit of life', since it is the smallest component capable of performing all of the characteristics of life. That is, cells can feed, generate energy, move, respond to stimuli, grow, excrete and reproduce. Our genes via their role in enzyme production are the controllers of cellular metabolism. Humans are complex organisms having cellular, tissue, organ and organ system levels of organization. Each level is interdependent and instrumental in sustaining the functions of life for the human body.

The characteristics of life are also interacting. For example, we must take in the raw materials of food and oxygen in order to provide energy, via the process of cellular respiration. This energy is needed to support metabolic reactions, such as those involved in preparing a patient for hospital admission, surgery, those accompanying trauma and surgery itself, and those necessary for the repair and growth of tissue generally, and post-surgery. Consequently, raw materials are referred to as the 'chemicals of life'. As a result of metabolic reactions waste products are generated and these must be excreted to prevent cellular disturbances.

Ultimately, every illness results from a cellular imbalance. The interdependence of the components of the body means that a failure of one function leads to a deterioration of others. This is reflected in the diverse signs and symptoms of ill-health which require clinical intervention to restore homeostasis. Homeostasis is arguably the most important concept within the academic discipline of physiology (Clancy and McVicar 1996). It represents the processes necessary for the maintenance of conditions under which cells, and hence the body, can function optimally; even small changes in body-fluid composition can disrupt biochemical activities within a cell and may even kill it. The interdependency of functioning at all levels of organization of the body means that such a disruption if not carefully monitored and controlled by the health-care team could be disastrous for the health of the patient.

The modern view that homeostasis is dependent on an integration of physiological functions is supported by Guyton (1987) who stated that *'Essentially all the organs of the body perform functions that help to maintain these constant conditions.'* Thus organ systems are homeostatic control mechanisms which regulate the intracellular environment throughout the body.

Principles of homeostasis

THE HOMEOSTATIC RANGE

Utilizing Cannon's (1932) definition of homeostasis as *'a condition which may vary, but remains relatively constant'*, and the normal ranges illustrated by patients' clinical chemistry charts, we have illustrated such physico-biochemical conditions graphically.[1] This graph is used throughout the accompanying textbook (Clancy and McVicar 1995) and this casebook as a model to explain:

1 homeostatic principles
2 homeostatic control system functioning
3 how the failure of control results in homeostatic imbalance which leads to illness
4 the principles of clinical interventions used to restore homeostasis.

The fluctuations in parameter value above and below the mean represent a homeostatic range within which the value is optimal; the minimum and maximum values of the range reflect slight variations within the individual as appropriate to daily routines (circadian patterns) and to account for individual variation within the population and according to the ageing process. The fluctuations occur as a normal phenomenon as a result of slight disturbances in equilibrium. If, however, the fluctuations are sufficient to cause a deviation outside the homeostatic range then parameter values begin to be suboptimal, resulting in a homeostatic 'imbalance'. Homeostatic control processes will normally then act to restore the equilibrium.

Occasionally, only one homeostatic control mechanism is present to redress the balance. For example, when blood-glucose concentration exceeds its homeostatic range then this results in the release of insulin, which promotes glucose utilization. More frequently, however, a number of controls are involved.[2]

All referals are from the main reference at the end of this case study.
[1]Refer to p.5, Figure 1.2.
[2]Refer to p.5, Figure 1.3.

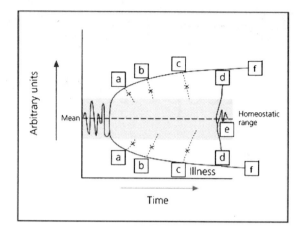

Figure 1.1 Clinical intervention following homeostatic control system failure. (a) Failure of short-term homeostatic control systems in re-establishing homeostasis. (b) Failure of intermediate-term homeostatic control systems in re-establishing homeostasis. (c) Failure of long-term homeostatic control systems in re-establishing homeostasis. (d) Clinical intervention to re-establish patient homeostasis. (e) Patient's re-established homeostasis. (f) Clinical intervention unable to re-establish patient homeostasis for e.g. terminal illness.

For example, when the blood hydrogen ion (H^+) concentration exceeds its homeostatic level, the homeostatic controls include intracellular and extracellular buffers, respiratory and urinary mechanisms. Numerous examples are provided in Clancy and McVicar (1995).

These corrective responses are time-dependent; whereas some respond quickly to the imbalance, their failure to re-establish homeostasis prompts other control mechanisms to correct the disturbance. The body therefore has many short-term, intermediate and long-term homeostatic control mechanisms. Using the above example, these are the buffers, respiratory and urinary mechanisms respectively.

The failure of control mechanisms results in the signs and symptoms of an illness which will be related to the homeostatic imbalances induced. Surgical, medical and nursing interventions employ various techniques and treatments to restore homeostasis. Some illnesses, such as terminal cancers, are not responsive to therapeutic intervention and consequently the imbalance results in long-term malfunction and eventually death (Figure 1.1). This connection of homeostasis to health care promoted the writing of this casebook.

HOMEOSTATIC FEEDBACK MECHANISMS

Most homeostatic control mechanisms operate on the principle of negative feedback. That is, when a homeostatic imbalance occurs, in-built and self-adjusting mechanisms come into effect which reverse the disturbance.

The initial imbalance in a physiological parameter is detected by sensory receptors. These receptors relay information about the imbalance to homeostatic control centres, which interpret the change as being above or below the homeostatic range, and the magnitude of the change. As a result they stimulate appropriate responses via effectors which bring about the correction of the imbalance by negative feedback, in order to restore homeostasis. Once the parameter has been normalized the response will cease. Thus, a failure of receptor response, control centre activity or effector organ, will prolong the imbalance and may be the cause of the illness which requires clinical intervention.[3]

There are times, however, when actually promoting a change, rather than negating it, is of benefit. Examples of such a positive feedback are observed during the blood-clotting process,[4] and oestrogen stimulating the leutinising hormone surge to promote ovulation.[5] Collectively such response maximizes the change required at that time.

Since such positive feedbacks induce change, the effects tend to be transient; most physiological systems utilize negative feedback mechanisms as a means of maintaining stability. An inability to promote change when necessary can, however, cause a change in health.

Physiological change can also be promoted by another process: that is, by altering the mean point and therefore the homeostatic range about which a parameter is regulated.

VARIATION OF THE MEAN

The mean or set-point reflects the importance of homeostatic control centres in the determination of the optimal value at which parameters are 'set'. Clearly a modulation of the set-point will either cause responses which promote a change in that parameter, or allow changes to occur uninhibited. In relation to the 'variation of the mean' in exercising muscles, the point could be emphasized that the intramuscular environment is maintained optimal by changes in blood pressure, cardiac output, etc., and also that whole body homeostasis (i.e. keeping blood pressure constant) would not then be appropriate. The 'closing down' of kidney function and gut function during exercise could be seen as a 'sacrifice' to maintain muscle homeostasis!

The capacity to modify set-points, therefore, is essential in certain circumstances and is of benefit, although the high temperature of a fever[6] may not make us appreciate it at the time! Set-point variation and positive feedback responses provide a flexibility to homeostatic processes. As with positive feedback responses, many of the changes promoted by set-point alteration relate to a specific situation and are short-lived. Some resettings are permanent, however, and so promote long-term change. These responses, for example, are vital to human development during the lifespan and allow for growth, functional maturation during fetal development and childhood, for pubertal changes during adolescence and even for the decline in functional capacity that accompanies old age![7]

[3]Refer to p.6, Figure 1.5.
[4]Refer to p.183–5.
[5]Refer to p.547, Figure 18.15.

[6]Refer to p.518.
[7]Refer to pp.561–93.

Reader activity

Using Chapters 1 and 2 in Clancy and McVicar (1995) *Physiology and anatomy: a homeostatic approach*, the reader could attempt to answer the following questions:

1 Describe how homeostatic negative feedback helps intracellular homeostasis.

2 Describe how organ systems, or homeostatic control systems, influence intracellular homeostasis.

3 Positive feedback is usually regarded as a homeostatic failure. Discuss this statement.

4 Give an example of a positive feedback mechanism in the body which does not result in illness.

5 All illnesses are ultimately a result of cellular imbalances. Discuss this statement.

6 Using the principles applied to the Operon theory, describe how:

(a) the levels of intracellular metabolites are controlled
(b) overexposure to UV light may cause skin cancers
(c) drugs may be used to correct homeostatic imbalances.

Main reference

Clancy, J. and McVicar, A. J. 1995 *Physiology and anatomy: a homeostatic approach*. London, Edward Arnold.

Other references

Clancy, J, and McVicar, A. J. 1996 Homeostasis – The key concept to physiological control. *British Journal of Theatre Nursing* 6 (2), 17–24.
Cannon, W. B. 1932 *The wisdom of the body*. New York, Norton.
Guyton, A. C. (1987). *Human physiology and mechanisms of disease*, 4th edition. London, W.B. Saunders.

The case of a man with hypothyroidism

Maggie Quinn

Learning objectives

1 To revise the physiology of the thyroid gland

2 To recognize the consequences of the failure of the thyroid gland's function

3 To revise the homeostatic mechanism for the maintenance of basal metabolic rate (BMR)

4 To understand the rationale for the nursing care of a patient with hypothyroidism

Case presentation

Edward is a 61-year-old man who during his working life had little reason to visit his doctor apart from routine health checks. Recently he has gained weight, despite a poor appetite. He complains of constantly feeling cold and has noticed that his finger nails have become brittle, and his skin appears to be thicker.

His family and friends have suggested that he visits his doctor as they have noticed that he is sleeping for longer periods during the day and he appears to have slowed in both his physical and mental reactions.

These observations prompted Edward to visit his GP who from the description of his symptoms suspected hypothyroidism. Serum tests confirmed this diagnosis, showing thyroid-stimulating hormone (TSH) levels of 4.5 μu/l (normal 0.15–3.2 μU/l) and free tetra-iodothyronine (T_4) of 9 pmols/l (normal 10–25 pmols/l).

Case background

Hypothyroidism is the result of a deficiency in the secretion of hormones produced by the thyroid. This may occur because of a primary deficiency of the thyroid itself, or because of a secondary deficiency occurring in the pituitary (i.e. a deficiency of thyroid-stimulating hormone) or in the hypothalamus (i.e. a deficiency of thyroid-stimulating hormone-release factor.) The thyroid gland secretes two important hormones: tri-iodothyronine (T_3) and tetra-iodothyronine (T_4) in order to maintain the basal metabolic rate.[8] The release of these hormones is determined by the presence of thyroid-stimulating hormone (TSH) from the anterior pituitary gland. High levels of T_3 and T_4 inhibit TSH release by the negative feedback mechanism.[9] In hypothyroidism T_3 and T_4 levels

All referals are from the main reference at the end of this case study.
[8]Refer to p.441:   [9]Refer to p.442, Figure 15.11.

are low and so TSH release is high. Thyroid hormones promote oxygen consumption and heat production and therefore impact on basal metabolic rate. The signs and symptoms experienced by Edward occur because:

- In hypothyroidism the basal metabolic rate may be decreased by up to 40 per cent of normal values and cold intolerance is commonly experienced. A slowing down of cognitive processes may result. Reduced basal metabolic rate also results in reduced cardiac output (observed as bradycardia), respiratory effort and blood pressure, leading to fatigue and weight gain.
- Serum lipids changes are due to altered cholesterol metabolism caused by depressed liver function.
- Hypothyroidism is often referred to as myxoedema. The latter is more correctly used to refer to the skin changes associated with chronic hypothyroidism. The skin becomes thicker because of the accumulation of mucopolysaccharides in the subcutaneous tissue and peripheral oedema is present. Tissue synthesis is reduced, resulting in flaky, dry skin, dry hair and brittle nails. Anaemia may be a symptom caused by reduced bone-marrow metabolism.

Care

The main objective of care is to restore Edward's normal metabolic rate by replacing the missing hormone (Figure 3.1). If thyroxine replacement treatment is successful the symptoms of myxoedema will be resolved. Careful monitoring of thyroxine replacements is required to avoid the symptoms of hyperthyroidism. Until treatment begins to take effect Edward's care must focus on the physical and mental effects of a reduced metabolic rate.

Edward may need help with his everyday activities because of the slowness of his physical and mental responses and fatigue. However, he should be encouraged to be self-caring wherever possible. The nurse and family must show tolerance towards Edward and allow him appropriate time to respond and carry out activities.

Edward may need protection from cold draughts and extra blankets should be provided for him if needed.

Edward is at risk of developing very dry skin and if necessary oil-based creams may be used to avoid this. The use of excessive amounts of soap should also be discouraged.

The nurse should assess Edward's nutritional requirements and liaise with the dietician if appropriate. A low calorie, high protein diet may be required to avoid the increase of weight. An increase in roughage may be needed to avoid constipation.

Explanations must be given to Edward and his family about his diagnosis and the effect it will have on his life. Symptoms of an overdose or underdose of thyroxine must also be given.

In summary, the rationale for care in hypothyroidism is to ensure that a normal metabolic rate is maintained by the use of hormone replacement. Monitoring of thyroxine levels must be undertaken to avoid hyperthyroidism.

Further information

Hypothyroidism is more prevalent among women than men and most commonly occurs/presents between the ages of 35 and 60. The majority of cases are caused by a dysfunction of the thyroid gland itself rather than a secondary cause. The most common cause of hypothyroidism in adults is an auto-immune condition known as Hashimoto's thyroiditis, although a significant number of cases occur because of an overeffective treatment of hyperthyroidism by surgery or radioiodine.

Main reference

Clancy, J. and McVicar, A. J. 1995 *Physiology and anatomy: a homeostatic approach*. London, Edward Arnold.

3 The case of a woman with hyperthyroidism

Maggie Quinn

> **Learning objectives**
>
> **1** To revise the physiology of the thyroid gland
>
> **2** To recognize the consequences of the failure of the thyroid gland
>
> **3** To revise the homeostatic mechanism for the maintenance of basal metabolic rate (BMR)
>
> **4** To understand the rationale for the nursing care of a patient with hyperthyroidism

Case presentation

Monica is a 52-year-old woman who has always been active and describes herself as 'living life to the full'. She has never had any serious illness.

However, two months ago she began to experience sleeplessness, 'hot flushes', nervousness and weight loss which she attributed to the onset of the menopause. She wondered if she was 'run down' because she thought she could feel swollen glands in her neck.

She made an appointment to see her general practitioner who on examination found her to have a pulse rate 98 and a blood pressure of 140/110 mm Hg. Her respirations were rapid and she exhibited signs of a fine tremor. The GP concluded that her glands were not swollen but that she had a goitre. He took blood samples in order to confirm a diagnosis of hyperthyroidism. Serum levels confirmed this, showing thyroid-stimulating hormone (TSH) to be suppressed (normal 0.15–3.2 mU/l) and free tetra-iodothyronine just within normal limits (normal 10–25 pmols/l).

Case background

Hyperthyroidism is a result of the oversecretion of thyroxine-based hormones by the thyroid gland (Figure 3.1). As with hypothyroidism this may be as a result of a primary problem of the thyroid itself or due to secondary problems associated with the release of TSH from the anterior pituitary gland. In hyperthyroidism, levels of tri-iodothyronine (T_3) and tetra-iodothyronine (T_4) are high while TSH is low because of subsequent negative feedback in the anterior pituitary gland. As a consequence of these changes the metabolic rate is increased. Monica experienced her signs and symptoms because:

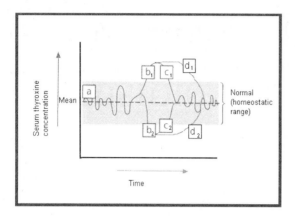

Figure 3.1 Hyperthyroidism/Hypothyroidism. (a) Thyroxine concentration representative of normal synthesis and release from the thyroid gland. The concentration fluctuates within the homeostatic range (i.e. optimal range) necessary for health. (b_1) Excessive thyroxine release. If normal regulation mechanisms fail, this will result in clinical hyperthyroidism. (b_2) Deficient thyroxine release. If normal regulation processes fail to correct this, clinical hypothyroidism will result. (c_1) Decreasing thyroxine concentration caused by normal negative feedback responses (i.e. decreasing TSH release from the pituitary gland). (c_2) Increasing thyroxine concentration caused by normal negative feedback responses (i.e. increasing TSH release from the pituitary gland). (d_1) Successful clinical intervention using anti-thyroid drugs, and/or partial or total thyroidectomy. (d_2) Successful clinical intervention using thyroxine treatment.

- In hyperthyroidism there is a general increase in the metabolic rate as a consequence of elevated T_3 and T_4 release. Increased catabolism and heat production result in an increased intolerance to heat. An increased demand for nutrients may lead to an increased appetite, since nutrients are used so rapidly the patient often presents with weight loss.
- The increase in metabolic activity leads to greater demands being placed on the cardiovascular system. Blood pressure rises and the depth and rate of respirations increase. The patient often presents with a tachycardia even when asleep.
- Other effects of excessive levels of T_3 and T_4 hormones in the body include those on the nervous system. Patients are often

irritable, display nervous energy and have difficulty sleeping.

Care

The aim of care is to restore Monica's normal basal metabolic rate. This involves a number of options for the medical staff. Most commonly the first line of treatment involves the administering of antithyroid drugs (Figure 3.1). These drugs interfere with the synthesis of thyroid hormones.

The patient may find it difficult to relax but the nurse will need to consider providing the patient with an appropriate environment for rest. Relaxation and rest must be encouraged to reduce stress on the cardiovascular system. A regular routine may help to avoid unnecessary disturbance and will ensure the patient understands what is expected of them. A cool environment may also help to promote rest.

The nurse should encourage the patient to take a high-protein, high-carbohydrate diet in order to meet the body's increased requirement for nutrients and to prevent tissue breakdown. The dietician should be involved in the care as it is not unusual for patients to require diets containing 4000–5000 kcal per day.

The increase in basal metabolic rate results in excessive heat production and perspiration. There is also an increased production of metabolic wastes, requiring dilution for elimination by the kidneys. Both these factors may lead to an increased fluid loss. Fluid replacement is vital and an intake of 3000–4000 mls per day is often required provided there are no contraindications such as cardiac dysfunction. Excessive tea and coffee drinking should be avoided, however, as caffeine will act as a stimulant.

The patient may need psychological support in order to cope with an altered body image and the impact of the disease on her lifestyle.

Explanations of treatment should be given to Monica and her relatives and the implications of the diagnosis for lifestyle discussed.

In summary: the aim of care in the case of hyperthyroidism is to restore the patient's normal metabolic rate. The monitoring of thyroid function should take place in order to prevent hypothyroidism.

Further information

Hyperthyroidism is far more common in women than men and more than 90 per cent of cases of hyperthyroidism are caused by the overactivity with the thyroid gland itself rather than any other secondary cause. This over-activity is caused by thyroid-stimulating auto-antibodies and the condition is often referred to as Graves disease. (Robert Graves was an Irish physician, 1796–1853).

While the first line of treatment is often with the use of medication, iodine which temporarily inhibits hormone release may also be used. In some cases surgical intervention is necessary, in the form of either a partial or total thyroidectomy, especially if the patient is deemed to be at risk of thyroid crisis. Thyroid crisis or thyroid storm is a life-threatening exacerbation of thyrotoxic symptoms. The patient is at risk from cardiac arrhythmias, particularly atrial fibrillation and shock. Treatment must be immediate in order to prevent death.

The long term complications of hyperthyroidism include muscle-wasting, osteoporosis, hair loss and exophthalmos. Many long-term effects of hyperthyroidism are irreversible, or only partially so, and patient education on the need for strict control is most important.

Conclusion

The maintenance of systemic bodily functions regulates appropriate cellular activities. These are determined by enzymes, the products of gene expression. The composition of the intracellular environment will influence the efficiency at which cells operate and, accordingly, it is regulated so as to be optimal. 'Homeostasis' refers to those processes which maintain the equilibrium, or balance, within the body compartments. Homeostatic control relies mainly on negative feedback mechanisms which act to reverse imbalances and to regulate the parameters close to the optimal mean value. The prevention of parameter variation can be detrimental under some circumstances. The promotion of change via positive feedback mechanisms or through a resetting of homeostatic set-points is then of benefit. The failure of negative feedback processes, of appropriate positive feedback responses, of set-point resetting, or a reduction in their efficiency, leads to illness. Surgical, medical and nursing interventions are largely concerned with supplementing normal anatomical, biochemical and hence physiological processes in order to re-establish the homeostatic status of the individual (re-examine Figures 1.1 and 3.1).

This section has considered two cases: hypothyroidism and hyperthyroidism, whereby intervention has been successful in maintaining a good quality of life for the patient. In other cases within this book, however, interventions are not quite so successful, for example, in rheumatoid arthritis and AIDS, where health care revolves around symptom management.

Adult case studies

The case of a woman with breast cancer

Sue Sides

Learning objectives

1 To revise the physiology of the female breast

2 To recognize the consequences of loss of control of the mechanisms which regulate cell division and the proliferation of breast tissue

3 To understand the basis for treatment of breast cancer

4 To understand the rationale for the nursing care of a woman with breast cancer

Case presentation

Janet is a 44-year-old married woman. Recently she discovered a painless lump in her right breast. Her GP referred her to a breast care surgeon who performed a needle biopsy and subsequently diagnosed an adenocarcinoma (adeno = of glandular tissue). Janet was admitted to hospital for a lumpectomy and excision of axillary lymph glands[1], after which she was referred to an oncologist for further treatment.

Janet has four children, all of whom she breast fed. There is no family history of breast cancer.

Case background

Breast cancer is the most common form of cancer occurring in women. Currently the UK has the highest mortality rate for this disease in Europe. The cause of breast cancer is unknown.[2]

Recently a gene has been identified which is implicated in familial breast cancer. However, only a small percentage of breast cancers show this inherited component and furthermore not all women with the gene actually develop the disease. Cancer-promoting genes are called oncogenes. Several oncogenes have been discovered which play a role in the development of breast cancer and it is thought that the cancer is likely to involve several genetic changes.

Hormones play an important role in the development of breast cancer, though their exact role is as yet unclear. Some studies link the long-term use of high-dose oral contraceptives with an increased incidence; others suggest that breast-feeding and the early age of first pregnancy offer some protection.

All referals are from the main reference at the end of this case study.
[1] Refer to pp.537–9 for revision of structure and function of the breast.

[2] Refer to pp.38–40 for a discussion of the aetiology of cancer.

Breast cancer represents a disorder of the mechanisms which regulate cell division and proliferation. Thus:

- Breast cancer cells lose many of the characteristics of normal cells.[3]
- There is an imbalance between cell-growth factors (produced by the expression of oncogenes) and those which suppress cell growth (produced by the expression of tumour suppressor genes).
- Cells divide at an accelerated rate, eventually giving rise to a palpable mass (tumour).
- Cancer cells are capable of migrating (metastasizing) from their site of origin to develop elsewhere. The axillary lymph nodes, liver, bone and brain are common sites of the metastatic spread for breast cancer. Hence the excision of axillary nodes at the time of lumpectomy.

Care

Treatment is aimed at the removal of the primary tumour and the eradication of any remaining abnormal cells which may lead to the development of metastases (see Figure 4.1). Combinations of surgery, radiotherapy, cytotoxic chemotherapy and hormonal treatment form the basis of treatment.

After an uneventful recovery from surgery, Janet was prescribed a course of radiotherapy followed by intravenous cytotoxic chemotherapy and hormonal therapy.

Radiotherapy is a local treatment, aimed at the destruction of any abnormal cells which may remain in the breast after the excision of the tumour. Radiation disrupts intracellular homeostasis by interacting with the atoms and molecules of tumour cells to produce irreversible damage and cell death.[4] A number of normal cells in the treatment area will also be affected but, unlike damaged tumour cells, normal cells can repair/reproduce themselves.

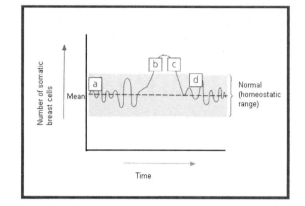

Figure 4.1 Breast cancer. (a) Numbers of somatic breast cells within homeostatic range, i.e. balance between production of cell-growth factors produced by oncogenes and growth inhibitors produced by tumour suppressor genes. (b) Abnormal breast tissue growth under the influence of oncogene expression and/or tumour suppressor gene inactivity. (c) Intervention to reduce number of abnormal cells (radiotherapy and/or cytotoxic chemotherapy) and suppress further abnormal growth (hormonal therapy). (d) Successful reduction of somatic breast-cell number. This is impossible to assess accurately and the individual will initially be considered to be in remission. That is, there are no visible signs of the tumour but there is a possibility that some tumour cells will have survived the therapies.

Cytotoxic chemotherapy is also aimed at the destruction of abnormal cells. As a systemic treatment, given intravenously, its side effects are more prevalent. Chemotherapy targets normal cells as well as malignant ones. White blood cells are particularly at risk,[5] therefore, blood-cell monitoring for the duration of treatment is essential. Chemotherapeutic drugs have a variety of actions but all in some way disrupt cellular growth, either by damaging DNA directly or by interfering with the ability of cells to utilize the substances necessary for growth and development.

Hormonal therapy manipulates the hormonal environment of breast-tissue cells to suppress cell growth. Oestrogens are an important factor in breast-cell activity. The drug

[3] Refer to pp.9–42 for the chapter on cell and cell function.
[4] Refer to p.45 for notes on radioisotopes.

[5] Refer to p.174, and pp.178–81 for information about white blood cells.

tamoxifen is an oestrogen antagonist: it binds to the oestrogen receptors of cells and thereby inhibits the growth-stimulatory effects of oestrogen and impedes the growth of malignant cells. Not all breast tumours are oestrogen-receptor positive, however. Although early trials suggest that tamoxifen therapy is likely to be beneficial in terms of increased survival, the degree of efficacy and optimal treatment regimens remains unclear, particularly in relation to those women who develop breast cancer before the menopause.

Janet will require skilled nursing care to enable her to cope with the demands of her treatment. Radiotherapy causes extreme fatigue and skin sensitivity, therefore the treatment is given in divided doses and careful observation of the treatment area is required. Although radiotherapy treatment is completely painless, it is often anxiety-producing. Patients may be overwhelmed by the physical environment of the radiotherapy unit, particularly the technical equipment. For the duration of each treatment the patient is also required to remain very still on the treatment couch and is alone, although staff are always visible behind protective lead screens (which do not permit the passage of radiation). Ideally each patient should visit the department before treatment is started to enable them to meet the staff and become familiar with the surroundings.

Cytotoxic chemotherapy is highly toxic, and often induces nausea and vomiting. These side effects can now be controlled effectively with anti-emetic drugs such as ondansetron. Janet's blood will be monitored at regular intervals for the duration of treatment to enable early diagnosis of leucocyte imbalances.

Patient support and education is an essential part of treatment. Not only will Janet be coming to terms with the diagnosis of a life-threatening disease but also recovering from surgery and facing several weeks of unpleasant treatment. Janet's family will also need support.

Further information

The most common types of breast cancer are infiltrating ductal adenocarcinoma (75 per cent and infiltrating lobular adenocarcinoma (5–10 per cent). The prognosis for both types is similar, with an overall 10-year survival rate of 50–60 per cent.

The outcome of breast-cancer treatment is largely dependent on the stage of the disease at diagnosis. Janet's disease was confined to her breast: i.e. there was no evidence of metastatic spread. There are, however, many factors which influence survival and individual biopsychosocial responses to treatment vary enormously.

The most common sites of metastatic spread from breast cancer are bone and liver: diagnosed by bone and liver scans respectively. Although not performed routinely, Janet may undergo these investigations at a later date, or should problems arise. Bone metastases usually present with pain in the affected area. Liver metastases are often asymptomatic until late in the course of the disease but may present with pain, liver enlargement (hepatomegaly) and/or jaundice.

Most hospitals now offer the services of specialist breast care nurses and there are also a number of charitable organizations and self-help groups which provide care and support for those with breast cancer. The British Association of Cancer United Patients (BACUP), for example, provides booklets, videos and has a telephone helpline for both patients and their relatives.

Main reference

Clancy, J. and McVicar, A. J. 1995 *Physiology and anatomy: a homeostatic approach*. London, Edward Arnold.

5 The case of a woman with climacteric syndrome

Louise Fuller

Learning objectives

1 To define the term, 'climacteric' and to identify the signs and symptoms that may accompany this lifestage

2 To understand why the climacteric can increase the risk of cardiovascular disease and osteoporosis

3 To identify the nurse's role in educating and assessing the client at this time

4 To understand the principles of hormone replacement therapy

Case presentation

Susan is a 52-year-old smoker, married with two grown-up sons. She runs her own business, buying and selling haberdashery products. Up until the last 12 months, Susan had loved her work and enjoyed a good social life, playing golf and swimming regularly. Then she had increasingly found she was getting tearful and moody, with feelings of inadequacy, and unable to cope. She was also experiencing frequent 'hot flushes', which she had found very embarrassing on more than one occasion. They were also occurring most nights with 'sweats' which meant her sleep was disturbed. After some discussion, Susan disclosed that she had also lost interest in what had been previously enjoyable sexual activity.

Susan is still menstruating, but irregularly, with long and short cycle lengths and sometimes heavy bleeding.

On examination, Susan's blood pressure was normal and she was of normal weight. Fasting lipids showed a slightly raised cholesterol and low-density lipoprotein and a lowered high-density lipoprotein. Full blood count, urea and electrolytes, liver function and thyroid function tests, were all within normal range. Serum follicle stimulating hormone (FSH) was 40 U/I (normal range 3.0–20 U/I) indicative of the climacteric.

Susan had a family history of coronary heart disease (CHD) with her mother having had a myocardial infarction (M.I.) at 60.

There were no outstanding social or psychological factors pertaining to these symptoms, so Susan's presentation reveals a case of climacteric syndrome, commonly termed the 'menopause'.

Case background

During a woman's normal reproductive phase, the relationship between the hypothalamus, the anterior pituitary and the ovary is homeostatically regulated within certain parameters.[1] As ovarian function diminishes, irregular periods and the symptoms described by Susan can occur, with some follicles requiring longer periods of FSH to stimulate them into maturity. Oestrogen (the predominant female hormone) levels begin to fall, which is detected by the hypothalamus. This increases the secretion of gonadotrophin-releasing hormone (GNRH), which in turn stimulates the pituitary to secrete more FSH (Figure 5.1). As the ovaries fail to respond, blood-serum levels of GNRH and FSH will rise, due to the breakdown in negative feedback. This transitional phase is known as the climacteric.[2] It denotes the terminal years of the reproductive phase, usually beginning in the woman's forties, and can last up to 10 years, culminating at the menopause (last period). It is therefore a physiological phenomenon.

The resulting oestrogen deficiency leads to numerous effects in the body, producing symptoms which are termed acute/intermediate and long-term.

In the acute/intermediate phase, which Susan appears to be experiencing, the main effects are of oestrogen withdrawal on the neuroendocrine system. This produces the symptoms of hot flushes, night sweats, insomnia, mood change and irritability. The lower urogenital tract produces symptoms of dyspareunia (pain on intercourse), loss of libido (to both of which Susan had admitted) and urethra syndrome.

Hot flushes are vasomotor in origin. Susan described a sensation of pressure in her head,

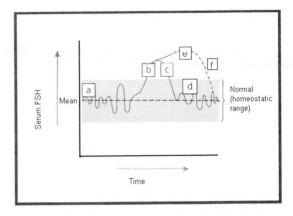

Figure 5.1 Climacteric syndrome. (a) Follicle-stimulating hormone (FSH) concentration within the homeostatic range necessary to promote follicle development during the menstrual cycle. (b) Elevated FSH concentration caused by excessive release from the anterior pituitary. (c) Decreased FSH concentration caused by negative feedback promoted by the initial oestrogen surge from the stimulated follicles. (d) Re-established FSH homeostasis. (e) Persistent elevation of FSH release as a result of a failure of negative feedback from inadequate oestrogen release at the onset of the climacteric. (f) Reduction of FSH release as a result of oestrogen-replacement therapy to restore the negative feedback mechanism.

followed by a feeling of heat travelling from her head down to her chest. It was associated with palpitations and sweats. Flushes, like this, are experienced by 70–80 per cent of women during the climacteric. It is the result of a disturbance in the function of the autonomic nervous system. The flushed face and chest are caused by cutaneous vasodilatation (adrenergic fibre activation[3]) and the sweating is caused by sympathetic cholinergic activation. There is a rise in skin temperature with an increased blood flow and a drop in body-core temperature.[4] The close proximity of the central thermoregulator to the GNRH containing neurones in the hypothalamus are thought to account for this inappropriate activity, believed to be caused by the high levels of GNRH.

All referals are from the main reference at the end of this case study.

[1] Refer to pp.545–50 for a discussion of the regulation of the female reproductive cycle.

[2] Refer to pp.583–4 for a discussion of the menopause.

[3] Refer to pp.389–91 for a discussion of the autonomic nervous system.

[4] Refer to p.517 and Figure 17.11 for an illustration of the regulation of body temperature.

Mood changes and feelings of anxiety and of being unable to cope are frequently experienced around the time of the menopause. The reasons for these psychological symptoms are not fully understood. Insomnia may be caused by the vasomotor symptoms, and can have an effect on mental processes and a general sense of wellbeing. Social class, culture-altered role, personality and preconceived ideas on this phase all affect how Susan will react.

Epithelia tissue requires oestrogen for normal growth and maintenance. The lining of the vagina (and distal urethra) atrophies, with the disappearance of the rugae and reduced blood flow causing sensory loss, and reduction of secretions leading to recurrent infections and vaginal dryness causing dyspareunia. This explains Susan's problems in her sexual relationship.

The long-term effects of oestrogen deficiency are on bone and the circulatory system. After years of ovarian failure, the arterial and skeletal systems are at risk and can lead to coronary heart disease, thrombosis[5] and osteoporosis.[6] The post-menopause period is a significant degeneration and the longer women live the longer women are at risk.

The mechanisms which affect the cardiovascular system that are contributory to coronary heart disease and thrombosis are:

- lipid metabolism (changes in blood lipoprotein concentrations)[7]
- carbohydrate metabolism with deterioration caused by insulin resistance[8]
- coagulation-factor changes[9]
- increased arterial tone causing increased arterial vessel resistance[10]

[5] Refer to pp.194–5 for a discussion of thrombosis.
[6] Refer to p.498 for notes on osteoporosis.
[7] Refer to p.105 and p.583–4 for a discussion of dietary lipids and lipid metabolism in menopause respectively.
[8] Refer to p.452 for a discussion of target-tissue responses in homeostasis.
[9] Refer to p.194 and pp.183–5 for a discussion of coagulation-factor changes and blood clotting respectively.
[10] Refer to pp.231–2 for a discussion of peripheral resistance.

These underlying changes, with Susan's family history and her smoking habit, put Susan at an increased risk of circulatory disease.

Oestrogen enhances the absorption of, and inhibits the excretion of, calcium by its indirect permissive action on calcitonin. Therefore falling levels of oestrogen with the ageing process have an effect on bone matrix and mineral content. The bones tend to become weaker and more brittle (osteoporosis) and are therefore more prone to fracture.

Care

At the initial consultation, the nurse's prime objective is to reassure Susan on this natural event and to discuss its effects and why her body is reacting as it is. At this stage, a great deal of information is given verbally and by giving Susan booklets to take home and read. This is an opportunity to assess lifestyle, and with history-taking and physical examination other causes of the symptoms can be eliminated.

A cervical smear and vaginal examination are carried out for evidence of abnormality. There are no such changes with the menopause, *per se*.

The treating of symptoms of oestrogen deficiency with oestrogen is logically inescapable, but it is contraindicated for those with cancer of the breast or uterus, unexplained vaginal bleeding and liver disease.

A mammogram carried out three months before was normal and Susan had no family history of breast cancer (cancer of the breast can depend on oestrogen). So there was no contraindication for hormone replacement therapy (HRT).

No plans for expensive bone densitometry were made as Susan had no obvious skeletal symptoms and although she smoked, she exercised regularly, ate well and had no family history of osteoporosis (all considered as low-risk for the development for osteoporosis).

The priority now for Susan's care is to:

• give symptom relief
• reduce the risk of cardiovascular disease
• prevent or delay osteoporosis

Careful explanation and reassurance about the nature of the symptoms and long-term effects were given to Susan. The treatment options were then discussed.

Treatment options can be divided into lifestyle advice (e.g. not smoking) non-hormonal therapy and hormone replacement therapy (HRT).

Before any decisions are made on treatment Susan was invited to return at a later date after she had had time to assimilate all the information given. Susan returned and had decided to try HRT.

The oestrogens used in HRT are natural and are manufactured to mimic 17-Betaoestrodial, oestriol and oestrone.[11] It can be taken in the form of tablets, creams, patches or an implant. Susan wished to try patches, where a patch containing oestrogen is applied to the skin and changed every three to four days. The benefits of transdermal administration are that systemic bioavailability for oestradiol via this route is high (85 per cent); it bypasses the liver and does not affect blood triglycerides.

Susan still has her uterus, therefore she also needs to take progestogen (synthetic progesterone) to protect the endometrium. In the normal menstrual cycle, the release of progesterone prevents the further growth of the endometrium and causes secretory transformation.[12] Oestrogen therapy would therefore cause overstimulation of the endometrium and could result in an abnormal build-up of endometrial cells (endometrial hyperplasia: a pre-cancerous condition) therefore progestogen is added in tablet form taken for 12 days every month. This causes a monthly breakthrough bleed similar to, but usually lighter than, a normal period.

[11] Refer to pp.432 and 445–6 for notes on endogenous oestrogens.
[12] Refer to p.549 for a discussion of the post-ovulatory phase of the menstrual cycle.

The side effects can be more upsetting than the climacteric symptoms and are usually caused by the progesterone component, i.e. nausea, breast tenderness, nipple sensitivity and leg cramps. These usually settle; if not, the medication can be changed to different doses or types. The compliance to continue with HRT relies on the expectations of Susan, symptom control and side effects. These are all fully discussed.

Susan is to be reviewed in three months' time, when symptom relief, side effects and any other problems can be assessed. Susan can also contact the nurse sooner if she feels the need.

Further information

The duration of treatment depends on the indication for symptom relief. A minimum of two to three years is usual, assessing symptoms when treatment is withdrawn. Epidemiological data about the duration for protection against arterial disease is limited. Once HRT is discontinued the benefits disappear and five to ten years' duration of therapy is usual. The oestrogen dose will be reduced gradually over two to three months to prevent the exacerbation of symptoms. The progestogen is continued until the oestrogen is discontinued.

The climacteric is a good example where parameters vary, as a failure of normal homeostatic control and disturbance in one system can exacerbate age-related changes in others.[13]

Main reference

Clancy, J, and McVicar, A. J. 1995 *Physiology and anatomy: a homeostatic approach*. London, Edward Arnold.

[13] Refer to pp.583–6 for a discussion of the ageing process.

6 The case of a man with a surgical wound

Gibson D'Cruz

Learning objectives

1 To understand the process that damaged or injured body tissues undergo in order that healing should take place

2 To understand how an alteration in the healing process may cause homeostatic imbalances in body-tissue function and structure

3 To acknowledge some of the factors that may affect the normal healing process

Case presentation

Mr James Jones is recovering, on a surgical ward, from surgery performed two days ago. The surgery was a removal of part of the stomach which had a cancerous growth contained within it. Today, James is receiving his nutritional requirements via a parenteral catheter that has been inserted into one of the central veins. His pain is being well controlled with an epidural infusion of diamorphine and this is enabling him to mobilize sufficiently.

James's other nursing problem is that the surgical wound must be healed before full recovery can take place. Today, the wound and the surrounding area is pink in colour, sensitive when a stimulus, such as touch, is applied, is slightly raised and feels warm when touched. There is a surgical dressing over the wound and when the dressing was last renewed, the dressing was discoloured by the impregnation of a wound discharge.

Case background

Broadly, the body is formed of four types of tissues that are structurally and functionally unique. These are epithelial, connective, nervous and muscular tissues. These tissues are prone to damage and injury and their function can only be regained following repair, regeneration and/or replacement of the damaged tissues.[1] An example of a homeostatic imbalance is seen when skin continuity is disrupted following trauma. One of the functions of the skin – a protective barrier between the internal and external environment – is not being achieved and the individual is more likely to allow bacteria into the body. This function is regained once healing has taken place (see Figure 6.1).

All referals are from the main reference at the end of this case study.
[1] Refer to pp.78–82 for a discussion of inflammation, and the repair and regeneration of wounds.

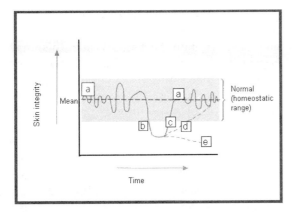

Figure 6.1 Surgical wound healing. (a) Normal structure and function of the skin. (b) Loss of skin integrity, e.g. as a result of trauma or surgical incision resulting in loss of function. (c) Commencement of the normal healing process by inflammatory response and repair/regeneration. (d) Delayed healing because of the presence of influencing factors, e.g. poor nutrition, infection, etc. (e) No healing as a result of influencing factors, e.g. ischaemia, dietary malnutrition.

Another method of classifying body tissues can be seen when their ability to repair is compared. Labile tissues, such as skin, are continually undergoing regeneration, and repair most speedily. Permanent tissues, such as nervous and muscular tissues, do not appear to repair following injury. In between these two are the stable tissues which, although they are able to repair in an anatomical sense, may not regain their functional ability. Where repair and regeneration takes place, the process is common to all body tissues. Skin is used here to illustrate this process as it is continually being damaged and the healing process can be visibly noticed.

Wound-healing involves two separate processes: inflammation, and repair with regeneration.

Inflammation

A series of steps are involved in the process of inflammation. The inflammatory response is the body's response to tissue damage which may or may not include bacterial invasion and is part of the non-specific local response.[2]

Inflammation begins with a release of many substances from a variety of cells including the injured cells.[3] The substances released include histamine, serotonin, prostaglandins and heparin. The effects of these chemicals are to increase capillary permeability and blood flow. A visual and tactile observation of the wound at this stage will reveal that the local temperature is raised and there is a reddening of the surrounding area; this will provide a reassurance that normal homeostatic responses have been activated following injury.

The next inflammatory process involves the removal of debris and micro-organisms. This activity is performed by phagocytes (microphages and macrophages).[4]

Repair and regeneration

Following the inflammatory process, cell repair and regeneration takes place. This process can be divided into three stages: migration, granulation and maturation.

This process usually begins with the clotting process leaving a residue in the form of a mesh. This mesh consists of fibrin that provides the support for the regenerating cells and later forms a scab, once healing is complete.

Cells now begin to migrate across the surface of the wound and this is the beginning of the creation of the new layer of cells. The reason for this migration of cells is not well understood although it is thought that there is a change in cell–cell communication that initiates the mobility of cells. There then follows the granulation stage of healing. New connective

---

[2] Refer to p.267 for a description of the appearance of the inflamed area.

[3] Refer to p.267 for an explanation of factors involved in the inflammatory response.

[4] Refer to pp.267–9 for a description of the stages of phagocytosis.

tissue cells – fibrocytes – are produced and the substances secreted by the macrophages, during the inflammatory process, facilitate this. These fibrocytes permeate through the fibrin mesh and collagen is secreted. This will ultimately form the scar of the wound.

The final stage of healing is the maturation of the wound. Here the collagen fibres begin to form an organized mesh and start to appear similar to the surrounding area in both its colour and texture. Fibrin, or the original blood clot, is removed by the action of a chemical called plasmin.

One important factor that ensures that the healing process is continued is the presence of an adequate blood supply. This supply is necessary to ensure that cellular functions such as mitosis take place.[5] Initially, a wound is usually devascularized at the time of injury. Angiogenesis – the creation of a new blood supply from the existing capillaries – provides a good blood supply and, consequently, the nutrients.

Care

Mr Jones's physiological function began to return and the parenteral nutrition was continued for the next three days. This was discontinued once he was able to consume a diet that contained sufficient nutrients and calories.

The wound dressing was changed every day as there was a discoloured fluid on the dressing. This discharge persisted for four days before it gradually began to reduce.

For the normal physiological process of healing to take place, two simultaneous measures are necessary. The first is that the process is not hindered by causing more trauma to the wound. This was achieved by applying an appropriate dressing that not only prevented this trauma but also provided a barrier to the entry of bacteria to the wound (see Figure 6.1). A result of the healing process

is that fluid containing both living and dead cells may be discharged from the wound. It may also contain bacteria. While this may appear to be an abnormality, it should be recognized that as this drainage is taking place, it is preventing an abscess, which is an accumulation of pus within a confined space, from forming.[6]

The second measure is that the healing process may be delayed by factors that may affect healing. The most important in Mr Jones's case was that of providing adequate nutrition, as certain nutrients, such as the vitamin B-complex, C and E and minerals such as zinc, are necessary for healing (see Figure 6.1).

During and after the healing process, two further events – wound contracture and scar contracture – may also take place. Wound contracture occurs during the healing process and is a normal homeostatic response as it serves to form a barrier against infection. Scar contracture, on the other hand, may be the result of the contracting of scar tissue and normally occurs after the wound has healed. Scar contracture usually results in loss of function. In extreme cases it can cause disfigurement.

Further information

In addition to the two measures cited above, other factors may also affect the healing process. These can be broadly divided into extrinsic and intrinsic to the wound. The extrinsic factors include pre-existing medical conditions, such as diabetes mellitus, that may alter the blood supply. Ischaemia is one factor that not only prevents healing from taking place at a normal rate, but may also mean that it does not occur. When body tissue does not heal, despite every attempt, tissue death may mean that physiological function also never returns to that area (see

---

[5] Refer to p.76 for notes on cell proliferation.

[6] Refer to p.83 for notes on pus, abscess, cyst and scar formation.

Figure 6.1). The other extrinsic factor that affects healing is age. Reduced cardiovascular function may also reduce blood flow into an affected site. The rate of cell division and cell metabolism is also reduced with advancing years.

Some of the intrinsic factors include the erosion or removal of granulating tissue when the dressing is changed. This may occur either when the dressing is removed or changed or when certain solutions are used to clean the wound.

When considering aspects of the healing of surgical wounds, it is worth remembering the distinction made by Hippocrates. He said that wounds either healed by first or second intention. In healing by first intention, as in Mr James's case, the edges of the wound are drawn together, placed in close approximation to its previous form and there is little or no tissue loss. The edges are held in place in some form, for example sutures, clips, etc. In healing by second intention, there may be substantial tissue loss and no attempt is made to draw the edges together. A good example of this is seen when there is either a sacral pressure sore or a leg ulcer. Healing takes place at the body's own rate but it may be enhanced with the use of appropriate dressings.

Main reference

Clancy, J. and McVicar, A. J. 1995 *Physiology and anatomy: a homeostatic approach*. London, Edward Arnold.

Further reading

Clancy, J. and McVicar, A. J. 1997 Wound healing: A series of homeostatic responses. *British Journal of Theatre Nursing* 7(4) 25–33.

7 The case of a young man with symptomatic HIV/AIDS

Judith Tyler

Learning objectives

1 To revise the specific immune response

2 To recognize the consequence of the inadequate immune system response which occurs with the destruction of helper T cells following exposure to infection by the human immunodeficiency virus (HIV)

3 To understand the complex interrelation of homeostatic functions whereby T-cell deficiency predisposes the affected individual to increased morbidity from a wide range of opportunistic infections

4 To understand the rationale for the symptomatic treatment of the presenting multiple pathologies exhibited by a person with acquired immune deficiency syndrome (AIDS)

5 To recognize that immunological competence currently cannot be restored in AIDS

6 To be aware of the range of care and support required by a person with AIDS

Case presentation

Timothy is 23 years old and was diagnosed as HIV-positive four years ago. Timothy can recall a period of general malaise and fever-like symptoms predating the confirmation of his HIV-positive status: a blood test which he underwent at the instigation of a former homosexual partner who was himself HIV-positive, and in the knowledge that his then lifestyle placed him at potential risk. By the age of 19, Timothy had had unprotected sex in a number of homosexual relationships, and for the last three years he has been in a loving and stable relationship with his partner, Jamie.

During the early part of this relationship, Timothy's health status changed from his being HIV-positive but asymptomatic, with the development of persistent generalized lymphadenopathy (PGL). Not only were his inguinal, cervical and axillary lymph nodes enlarged, but throughout this time he noticed a recurrence of the initial fever-like symptoms: at times his temperature was elevated, he perspired excessively (particularly experiencing 'night sweats') and he suffered debilitating bouts of persistent diarrhoea and nausea. Oral candidiasis ('thrush') has further diminished Timothy's enjoyment of his food, causing malnutrition and a weight loss (cachexia) from 68.66 Kg to 50.9 Kg.

Timothy has suffered from chronic fatigue, lethargy and quite severe mood swings. Over the preceding six months, he has encountered significant visual impairment caused by cytomegalovirus (CMV) retinochoroiditis. Within the last three weeks he has developed the symptoms of a severe chest infection which has resulted in his readmission to a local AIDS unit with a diagnosis of pneumocystis carinii pneumonia (PCP). He has been profoundly anxious and depressed at the inexorable progression of his disease and the effect that he sees this having on his partner, family and friends.

Timothy's disease management to date has consisted of antiviral medication (zidovudine/AZT), nutritional supplements and symptomatic treatment for a range of opportunistic infections as these have occurred, with social support, the use of some complementary therapies (massage, aroma therapy and relaxation techniques) and counselling, provided through a system of multi-agency cooperation and AIDS charity networks.

Case background

Innate short-lived or natural passive immunity is conferred on the neonate prenatally and through initial lactation by the mother, but by the time the child is three months old it needs to begin to acquire its own immunity which will enable it to produce a specific immune response against each invading pathogen.[1] The active agents of acquired immunity are the lymphocytes, with B-lymphocytes responsible for humoral immunity and T-lymphocytes for cell-mediated immunity.[2]

Acquired immune deficiency syndrome (AIDS) may be seen predominantly as a dysfunction of cell-mediated immunity. It is a form of secondary immune dysfunction caused by infection with the retrovirus designated as the human immunodeficiency virus. HIV is a retrovirus (it contains RNA, not DNA,[3] and utilizes reverse transcriptase enzyme to convert the RNA to DNA within an infected cell), with a particular affinity for the helper T-cells which it impairs and destroys, also impairing the function of cytotoxic and lymphokine-producing T-cells and compromising the efficiency of the B-lymphocytes. The resultant undifferentiated antibody production means that people with AIDS are unable to develop specific humoral immunity to any new antigens. This, combined with the massive reduction of helper T-cells and the probable impairment of those which survive, produces a terminal immunological deficiency, shown diagrammatically in the homeostatic graph (see Figure 7.1), and exposes the individual to repeated opportunistic infections (which may be viral, bacterial or fungal in origin). Collectively, the symptoms produce the characteristic presentation of AIDS.[4]

Care

The initial diagnosis of Timothy's HIV status would probably be made at a clinic for genito-urinary medicine (GUM clinic), following pre-test counselling from a specialist nurse adviser. This would be followed by post-test counselling and support, with attendance at

All referals are from the main reference at the end of this case study.
[1] Refer to pp.269–70 for an outline of the specific immune response.

[2] Refer to pp.270–80 for a more detailed discussion of the specific immune response.
[3] Refer to pp.29–33 for a description of the structure of DNA and RNA, and their role in protein synthesis.
[4] Refer to pp.286–7 for notes on the signs and symptoms of AIDS.

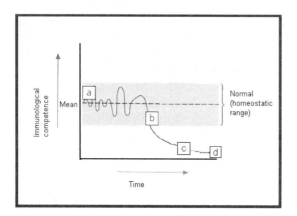

Figure 7.1 HIV and AIDS. (a) Normal immunological competence before HIV infection. (b) Potential immune response compromised by HIV infection. (c) Hyposensitive immune response occurring with symptomatic HIV and AIDS. (d) Terminal acquired immune deficiency syndrome: AIDS (i.e. death caused by an opportunistic infection e.g. pneumocystis carinii pneumonia, kaposi sarcoma, etc.

the GUM clinic for much of this medical care. Once symptomatic HIV presents, the GP and the primary health-care team will become more fully involved, with community nursing input during acute episodes not requiring hospital admission.

Ideally, intervention would seek to destroy HIV and restore homeostasis. In reality intervention has two general aims:

1 to slow the progression of HIV infection

2 to manage conditions arising from opportunistic infections.

The initial diagnosis of Timothy's HIV-positive but asymptomatic status would not be accompanied by any medical intervention, but when the condition known as PGL was recognized, he would commence continuing treatment with the specific anti-viral drug zidovudine (AZT), which inhibits reverse transcriptase (enzyme). He would also receive a short course of acyclovir (zirovax) to reduce the risk of an early reactivation of the herpes simplex virus.

This combination of drugs has been shown to have some effect in delaying the progression of HIV infection and may even delay the onset of AIDS.

The persistent diarrhoea is variously managed through the use of anti-diarrhoea drugs: loperamide hydrochloride (imodium), diphenoxylate hydrochloride with atropine sulphate (lomotil) and codeine phosphate. Replacement fluids such as dioralyte will help to redress fluid and electrolyte loss arising from the diarrhoea.

Outbreaks of oral candidiasis are contained by the topical application of an anti-fungal drug (nystatin) with the use of acyclovir and proprietary compounds to treat the sores around Timothy's mouth and cracked lips.

In the early symptomatic stage of the disease, Timothy would be referred to a dietician for advice, support and access to an appropriate range of nutritional supplements.

Timothy's visual problems are attributed to infection from a previously latent source of cytomegalovirus (CMV) reactivated by HIV infection. CMV commonly occurs within the wider community, as evidenced by the 50 per cent of women tested who are of reproductive age and who are found to have the antibody (Pratt 1991).

In the complex pathology of AIDS, it is often difficult to differentiate those symptoms which result from reactivated CMV infection, but it may be one of the contributory organisms involved in the production of gastrointestinal disorders, nausea, ulceration and diarrhoea. CMV retinochoroiditis results in the progressive and often rapid deterioration of sight, and possible ultimate blindness. A specialist nurse within the community nursing team will administer an intensive intravenous course of the anti-viral drug ganciclovir, reducing to a maintenance dose, to delay the progression of the condition. This treatment regime, however, is seldom successful in achieving more than a temporary alleviation of symptoms. Specialist equipment for use by the partially sighted may be acquired from the Royal National Institute for the Blind

(RNIB), possibly with assistance from AIDS charities.

Pneumocystosis (PCP) begins insidiously with a troublesome dry cough and some chest pain, particularly noticeable on inspiration. On admission Timothy would be febrile, cyanosed and in acute respiratory distress (see Clancy and McVicar 1997 for a discussion of pneumonia). His immediate management required symptom relief and an accurate differential diagnosis to exclude other types of pneumonia. Sputum obtained for culture and sensitivities, with the results of the chest X-ray, will be used to confirm the diagnosis of PCP, hopefully without the need for bronchoscopy and biopsy.

High-dosage intravenous co-trimoxazole (Septrin) is the drug of choice for PCP, administered with an antihistamine to counteract the familiar side-effects of rash and itching.

While universal safety precautions are observed by all participating staff, particular care and communication is required to prevent the severity of Timothy's chest infection and his impaired vision from compounding his sense of isolation and despair. His partner and his family, to whatever extent they are able, should be encouraged to participate in Timothy's care and support. It may be that just by reading to him they will feel able to contribute to his care, thus helping themselves to come to terms with the situation and providing him some pleasure and involvement.

The prognosis for Timothy now that PCP has been diagnosed is not good. It is imperative to discuss with him and those whom he chooses to involve, his future disease management, any respite care which may be afforded (often through the support of an AIDS charity), the continuation of complementary therapies as appropriate, and arrangements for his eventual terminal care.

The homeostatic deficit demonstrated in an individual with AIDS is progressive and irrevocable. In the present state of knowledge, there is no way that immunological competence can be restored (see (d), Figure 7.2).

The need for continued support and counselling of all those who are affected by Timothy's illness cannot be underestimated.

Further information

HIV is a blood-borne virus found in body fluids, whose principal mode of transmission is through sexual activity. In the developed world, despite some increase in heterosexual transmission, and decrease in homosexual transmission, this continues to be more commonly male-to-male.

HIV infection occurs as the invading virus becomes bound to the host cell receptor. Here the vesicle carrying the virus is able to release it into the cytoplasm of the host cell where the structure of the virus disintegrates, freeing both the RNA and the reverse transcriptase enzymes which are used to copy the RNA to produce a DNA molecule compatible with incorporation into the host DNA (the distinguishing characteristic of a retrovirus).

The infected DNA enters the nucleus of the host cell to become permanently incorporated into the host DNA, forming a provirus. Viral replication occurs, and the new viruses are forced out of the host cell into the blood stream, utilizing some of the host cellular material to achieve a form in which they are able to target their own host-cell receptors.

A long incubation period follows, although some patients will initially experience an acute illness similar in presentation to glandular fever. Antibodies to HIV are generally produced within a three to six month period (the 'window period' before which the infected individual undergoes seroconversion, and will then test HIV-positive).

The ultimate progression of the disease is determined by a number of existing factors: the existence of any latent viral infection, other defects or aspects which further compromise the immune system, and what is thought to constitute a genetic predisposition to the development of the various symptoms which together form the syndrome.

Main reference

Clancy, J. and McVicar, A. J. 1995 *Physiology and anatomy: a homeostatic approach*. London, Edward Arnold.

Other references

Clancy, J. and McVicar, A. J. 1997 Theatre hypoxia as a failure of respiratory homeostasis. *British Journal of Nursing* 6 (10) 15–22.

Pratt, R. J. 1991 *AIDS: a strategy for nursing care*, 3rd edn. London, Edward Arnold.

The case of a man with chest pain: myocardial infarction

8

Julia Hubbard

Learning objectives

1 To explain the possible causes of myocardial infarction and the patho-physiological changes that occur within the coronary arteries

2 To demonstrate an understanding of pharmacological intervention in the treatment of acute myocardial infarction

3 To explain the rationale for care of a person with acute myocardial infarction

4 To identify the health education needs in the on-going management of individuals with ischaemic heart disease

Case presentation

John is 54 years old. He is a slightly overweight builder who smokes 20 cigarettes a day and takes little recreational exercise. Today John was admitted to the coronary care unit with a two-hour history of central crushing chest pain which radiated to his left arm and jaw. The pain occurred at rest and was accompanied by breathlessness, nausea and vomiting. On examination John was found to have severe chest pain. He was cold and clammy to touch, tachycardic and hypotensive.

Following a rapid physical examination, a 12-lead electrocardiogram demonstrated convex ST segment elevation[1] over the anterior chest leads (v1–v6). As these electrocardiogram changes are indicative of an acute anterior myocardial infarction, blood tests for the cardiac specific enzymes, creatine kinase isoenzyme CKMB and Troponin T, were taken to substantiate the diagnosis.

Case background

Myocardial infarction is a homeostatic imbalance which reflects myocardial oxygen insufficiency

All referals are from the main reference at the end of this case study.
[1] Refer to pp.212–13 for a discussion of ECG.

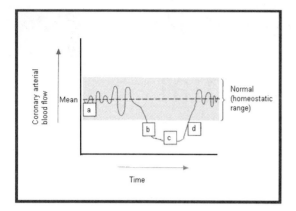

Figure 8.1 Myocardial infarction. (a). 'Normal' blood flow in coronary arteries. (b) Coronary arterial insufficiency caused by stenosis or vasospasm and ensuing myocardial ischaemia. (c) Partial or complete artery occlusion and subsequent myocardial infarction. (d) Restoration of near normal cardiac function following reperfusion with thrombolytic therapy.

(myocardial ischaemia) usually arising from an occluded coronary artery (Figure 8.1). Blood flow to the tissue beyond the occluded site ceases; if the occlusion is complete the myocardial tissue dies (myocardial necrosis).[2] Occlusion is usually due to atheroma and subsequent thrombosis formation (Pepine 1989).

Myocardial infarction is commonly associated with severe chest pain. Insufficient oxygen supply to the myocardium inhibits the complete metabolism of glucose for energy, and as a result anaerobic metabolism occurs and lactic acid accumulates. Lactic acid is known to stimulate pain fibres found within the myocardium. The origin of the chest pain is therefore the myocardium itself: the impulses travel from the myocardium via sympathetic fibres to the thoracic sympathetic ganglia to nerve roots T1–T5.[3] These spinal nerves supply the anterior chest wall and the inner aspect of the arm and head. For that reason, pain is felt in the region bounded by these thoracic nerves, including the left arm.

[2] Refer to pp.252–3 for a discussion of ischaemic heart disease.
[3] Refer to p.352 and to Figure 13.4 for a diagram of the spinal nerves.

A major myocardial infarction may result in death from cardiogenic shock, cardiac dysrhythmias or cardiac rupture.[4] If the individual recovers, their lifestyle may be restricted by chronic left ventricular failure or angina pectoris.

The risk of suffering a myocardial infarction in a given individual or community reflects the interplay between genetic susceptibility to the disease and environmental factors such as smoking, elevated cholesterol levels and a lack of physical exercise.[3] There are also gender differences in relation to heart disease, with the incidence being highest in men and post-menopausal women (Tsunuda 1996).

Care

Treatment for acute myocardial infarction focuses on the recanalization of the occluded artery. This limits the size of the infarction (necrotic area) by reperfusing the ischaemic zone before hypoxic injury escalates to necrosis.

Immediately following admission John had an intravenous cannula inserted and diamorphine 5mg and maxalon 10mg IV was administered for pain relief and persistent nausea. Diamorphine acts quickest, and has the added benefits of reducing anxiety, ensuring rest and reducing myocardial preload caused by venous pooling (subcutaneous/intramuscular administration is less reliable and may cause bleeding with subsequent thrombolysis). Sublingual or intravenous nitrates, thrombolytics and intravenous beta blockers may also relieve cardiac chest pain. During the acute phase of John's infarction he was nursed in bed in a semi-recumbent position and attached to continuous cardiac monitoring for constant observation of his heart rate and rhythm.

Once John's diagnosis had been confirmed by the cardiologist he was prescribed aspirin

[4] Refer to pp.249–50 for notes on shock.

150 mg orally and streptokinase 1.5 mega units intravenously over one hour. Throughout this time a nurse remained with John, monitoring his blood pressure and observing for signs of anaphylaxis.

Research shows that intravenous thrombolysis reduces mortality in patients with acute myocardial infarction (ISIS 3 1992) (Figure 8.1). Three thrombolytic agents currently available in the United Kingdom are streptokinase, alteplase and anistreplase.[5] The best results in terms of both the preservation of ventricular function and improved survival are obtained by starting thrombolytic therapy as soon as possible after the onset of symptoms (ISIS 2 1988). Conversely, the potential benefit of thrombolytic therapy in a patient with a small myocardial infarction who is pain-free and haemodynamically stable has to be weighed against the risk of a major bleeding complication (1 per cent) and sensitization to streptokinase (ISIS 3 1992). Aspirin has also been proved to enhance the benefit of thrombolysis (ISIS 2 1988).

Following 48 hours without ischaemic chest pain, John was transferred to a medical ward for mobilization and rehabilitation. During the rehabilitation phase of hospitalization particular attention is paid to modifying an individual's cardiac risk factors in the hope of preventing recurrent infarction. In John's case he would need advice on stopping smoking, diet, weight loss and the importance of exercise.

Once acute myocardial infarction has been diagnosed, the aims are to abolish the symptoms and to restore normal or best possible long-term cardiac function.

Further information

The diagnosis of acute myocardial infarction is made essentially in three steps: the patient's history, the electrocardiogram, and enzymes' studies. In many ways the patient's story of his illness is the prime factor in reaching the diagnosis and admitting the patient to hospital. However, the history, regardless of how typical it may be, is not diagnostic in its own right and other steps must be taken to prove the acute infarction has actually occurred.

The electrocardiogram is the single most valuable immediate diagnostic tool. However, additional confirmation of the diagnosis can be made by detecting the raised plasma activities of the cardiac enzymes (intracellular enzymes which leak from injured myocardial cells into the blood stream) particular attention is paid to CKMB and Tropinion T. Retrospective diagnosis can also be made from asparate transaminase (AST) and lactic dehydrogenase (LDH). However, these enzymes are not cardiac specific.

Continuous cardiac monitoring is essential. During the early stages of acute myocardial infarction there is autonomic stimulation with subsequent increased catecholamine secretion. This, in conjunction with ischaemic myocardium, may lead to death from cardiac dysrhythmias, particularly ventricular fibrillation.

Prognosis in myocardial infarction is principally related to the age of the patient and the residual left-ventricular function. Health education is an extremely important part of post-infarction care with particular attention paid to modifiable cardiovascular risk factors. Information should be made available on smoking, diet, weight and exercise (including sexual activity).

The combination of pharmacological and educational interventions should lead to the effective long-term management of ischaemic heart disease.

Main reference

Clancy, J. and McVicar, A. J. 1995 *Physiology and anatomy: a homeostatic approach*. London, Edward Arnold.

[5] Refer to p.183 and to Figure 8.9 for revision of normal clot lysis.

Other references

ISIS 2 1988 Second international study of infarct survival. *The Lancet* 13 August, 349–60.

ISIS 3 1992 Third international study of infarct survival. *The Lancet* 28 March, 753–66.

Pepine, C. J. 1989 New concepts in the pathophysiology of acute myocardial infarction. *The American Journal of Cardiology.* July, 64, 2B–7B.

Tsunuda, D. 1996 Acute myocardial infarction. *American Journal of Nursing* 96 (5), 38–9.

Further reading

Clancy, J. and McVicar, A. J. 1996 Shock: a failure to maintain cardiovascular homeostasis. *British Journal of Theatre Nursing* 6 (6) 19–25.

The case of a man with hyperkalaemia

9

Gibson D'Cruz

Learning objectives

1 To revise the role of potassium in maintaining cardiac function

2 To understand the rationale behind strategies in the management of hyperkalaemia

3 To appreciate the importance of safely administering potassium supplement in order to prevent hyperkalaemia

---

Case presentation

John is a 48-year-old man who had undergone abdominal surgery a day earlier. The surgery performed was the resection of a growth in the ileum and an anastomosis was created to restore continuity of the small bowel. At present, he is receiving parenteral nutrition through a catheter into a central vein. John also has a peripheral venous cannula for infusions to correct dehydration and provide access for the administration of drugs. Five hundred mls of normal saline is being administered here at a rate of 150 mls per hour.

The profile of John's biochemistry assay, from a sample of venous blood taken today, showed that the potassium level was 3.0 mmol/litre (normal range 3.8–5.1 mmol/litre.[1] In order to correct this hypokalaemia, 20 mmols of potassium was prescribed. This potassium supplement was diluted in the

All referals are from the main reference at the end of this case study.
[1] Refer to pp.340–1 for a discussion of the regulation of potassium concentration.

normal saline. This fluid was begun at 1.00 pm and was expected to be infused by 7.00 pm.

At 1.20 pm, John was noted to have a slow pulse rate, muscle weakness and facial paralysis. Scrutiny of the peripheral infusion showed that the litre of fluid had been administered.

An electrocardiogram (ECG) recording demonstrated a second degree (Mobitz type I) heart block with an elevated T-wave[2] and hyperkalaemia was diagnosed. A venous sample of blood, analysed for potassium levels, later confirmed the diagnosis. The serum potassium was 7.0 mmols/litre and there was also a mild acidosis.

Case background

During and after episodes of rapid fluid loss, starvation or gastro-intestinal disturbances such as diarrhoea, vomiting and abdominal fistulae, serum potassium levels are likely to be reduced. This scenario is likely to have occurred in John's case where two of the

[2] Refer to pp.212–13 for a discussion of the ECG.

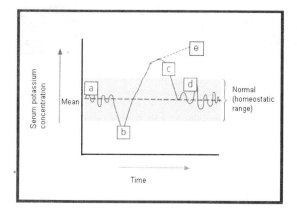

Figure 9.1 Hyperkalaemia. (a) Serum concentration of K⁺ within homeostatic range. (b) Reduced serum concentration of K⁺ caused by reduced dietary intake. (c) Elevated serum concentration of K⁺ caused by rapid infusion of potassium supplements. (d) Return of serum concentration of K⁺ within homeostatic range following management of hyperkalaemia. (e) Serum concentration of K⁺ elevated too much, too quickly, producing cardiac arrythmias and potentially cardiac arrest.

factors that may cause hypokalaemia – surgery and the need to restrict a normal diet – were present. There was no evidence of the presence of an intestinal fistula.

A speedy and efficient method of correcting hypokalaemia is through the use of potassium supplements (Figure 9.1). These can either be administered orally or intravenously. However, an excessively high potassium concentration can be a potential hazard if the administration of potassium supplement occurs too rapidly.

Care should be taken when administering potassium. If given intravenously, the correct dilution should be adhered to and the speed at which it is given must also be closely monitored. The recommended dilution is an ampoule (10 mls) containing 20 mmols of K⁺ which can be added to 500 mls of sodium chloride or glucose solutions to create a solution containing 40 mmols per litre. This solution should be administered over three hours (British National Formulary 1996).

Potassium serves three main functions in maintaining homeostasis:

- it helps neuromuscular functioning
- it helps metabolic activities such as protein synthesis
- it constitutes part of the body's buffer system, maintaining the pH of the body fluids as a short-term homeostatic control

A rapid shift in serum and/or intracellular potassium levels (i.e. a change in the potassium concentration gradient across cell membranes) can have an effect on a cell's resting potential[3] and so have consequences for excitable cells such as neurones and muscle fibres. A reduced serum potassium level, or high intracellular potassium, may cause an elevation in the resting potential thus making cells, such as cardiac cells, more difficult to stimulate. This hyperpolarization has an effect on the cardiac muscle, and the consequent heartbeat is slow and feeble. A raised serum-potassium concentration, on the other hand, initially lowers the resting potential, making cells more excitable. If sufficient, however, the elevated potassium concentration prevents the repolarization of the ventricles from taking place. This may cause severe cardiac arrhythmias and, if untreated, may lead to complete heart block and consequently to cardiac arrest (Figure 9.1).

Another effect of the low potassium level relates to a disruption in body fluid pH.[4] A movement of hydrogen ions (H⁺) out of a cell is normally compensated by a movement of potassium ions (K⁺) in the opposite direction to maintain positive ion balance (Marieb 1992). In episodes of hypokalaemia, there is likely to be a shift of potassium ions out of cells in order to maintain a normal concentration gradient across the cell membrane. This may lead to alkalosis as H⁺ pass into the cells to compensate. During hyperkalaemia, the reverse happens and this causes acidosis.

As the signs and symptoms that John was experiencing are not unique to hyperkalaemia,

---

[3] Refer to pp.361–2 for an analysis of the membrane potential.

[4] Refer to pp.87–9 for a discussion of acids and bases.

two investigations are necessary to confirm this diagnosis. The first is an ECG and this will show the signs of heart block. In second-degree heart block, the ventricles do not receive the impulse from the AV node at the appropriate time for the single depolarization to take. The ventricles may eventually continue to beat at their own, but weak, rate. This rate is not sufficient to maintain total body circulation and is life-threatening.

The second technique involves measuring K^+ in a venous blood sample. Although accurate, this must at times be treated with caution. It may take some time for changes in potassium levels to be detected in the venous sample, therefore greater attention must be placed on ECG recordings for accurate diagnosis.

Care

The main aim of the interventions for hyperkalaemia is to decrease extracellular potassium.

The first, but important, step is to ensure that no further potassium is being administered. This action was not needed for John because the infusion of normal saline with potassium had already been completed. The fluid now administered via the peripheral cannula was 5 per cent dextrose.

The next stage in the treatment was the administration of a mixture of 20 per cent glucose containing 20 units of actrapid insulin. The combination is such that the insulin promotes the utilization of the glucose, but also promotes the uptake of potassium by muscle and liver cells. This was administered over one hour.

Gradually John's pulse became stronger and the ECG recording showed a return to normal sinus rhythm. A sample of venous serum, taken at 2.30 pm, showed that the serum-potassium level was extremely close to its homeostatic range (5.2 mmols/litre).

At the end of this infusion, John's peripheral line was maintained with dextrose saline, a litre being administered over the next 12 hours.

Over the next few days, John's intestinal function began to return and he was commenced on a small but palatable diet. The parenteral infusion was discontinued and no further electrolyte imbalances were noted.

Further information

There are other causes, apart from the inappropriate management of potassium supplements, for hyperkalaemia to develop. One other cause is renal failure[5] where the body's ability to excrete fluid and electrolytes in sufficient amounts is impaired. Acute renal failure, secondary to surgery, can be a complicating factor in surgical care, though here this was not the case.

Another cause of a disruption of the homeostatic balance of potassium is the action of diuretics. Some of these drugs, such as amiloride, do not excrete potassium during diuresis, while others such as frusemide, do. It is therefore important that when patients are administered diuretics, their electrolyte balance is monitored.

There are three other methods of reducing serum potassium levels.

First, a cationic exchange. Drugs such as calcium resonium or resonium A are given either orally or, more preferably, rectally. The sodium ions contained within it are exchanged for potassium. The site at which this exchange occurs most is in the large bowel: hence the preference for the rectal administration. This route of administration must be used with care because in certain situations, as in this case following intestinal surgery, the rate of absorption may be altered because of a paralytic ileus. It is possible to remove 1 mmol of potassium from the extracellular fluid with the administration of 1 gram of either of these drugs.

[5] Refer to pp.345–6 for an outline of renal failure.

Second, a change in the body pH. As a result of hyperkalaemia, acidosis may be present for the reasons suggested earlier. The administration of an alkaline substance, such as sodium bicarbonate, may be effective in managing this.. By reducing the acid balance, potassium is shifted into the cells and this brings about both a reduction in serum potassium and a decrease in acid levels.

Third, dialysis. This is a very effective mechanism but one that is used as a measure of last resort. Dialysis maintains body-fluid composition within a homeostatic range but this technique is one that requires the insertion of arterial and venous lines for access and is also expensive.

Main reference

Clancy, J. and McVicar, A. J. 1995 *Physiology and anatomy: a homeostatic approach*. London, Edward Arnold.

Other references

British National Formulary 1996 *Number 31*. British Medical Association and Royal Pharmaceutical Society of Great Britain (publishers).

Marieb, E. N. 1992 *Human anatomy and physiology*. California, The Benjamin and Cummings Publications Company Inc.

The case of a woman with rheumatoid arthritis

10

Janice Mooney

Learning objectives

1 To revise the structure and function of joints

2 To revise the auto-immune/inflammatory response

3 To be aware of the care provided and the multidisciplinary management of the disease

4 To understand the rationale and possible side-effects of disease-modifying anti-rheumatic drugs

Case presentation

Mary is a 50-year-old married woman who works as a shop assistant. She visited her general practitioner complaining of extreme tiredness together with painful, stiff, swollen hands and knees, particularly first thing in the morning. She is unable to dress herself without help and has become limited in her walking ability. This has made carrying out the activities of daily living very difficult. Previously she was completely independent.

Non-steroidal anti-inflammatory drugs did not relieve her symptoms. Routine blood tests revealed that she was anaemic, Hb 9.8 g/dl with a normocytic, normochromic picture[1] and she had an elevated erythrocyte sedimentary rate (ESR) at 90 mm/hr and a positive rheumatoid factor test with a titre of 1280. She was referred to the rheumatology department.

Examination confirmed synovitis of both wrists, metacarpophalangeal and proximal interphalangeal (the metacarpal and phalanges are bones of the hand joints). Effusions of both knees and restricted shoulder movements were detected.

On questioning she had lost three kilos in weight and early morning joint stiffness lasted four to six hours. A diagnosis of rheumatoid arthritis was made.

Case background

Rheumatoid arthritis is a chronic, debilitating, inflammatory arthritis for which there is no cure. It is more common in females than males

All referals are from the main reference at the end of this case study.
[1] Refer to pp.172–82 for a discussion of blood cells.

(3:1). It involves inflammation of the synovium[2] that lines both joints and tendon sheaths, causing pain, stiffness and swelling, which can cause long-term joint damage, leading to decreased mobility. It is characterized by periods of remission and flares, and may produce tiredness, anaemia and weight loss. Extra-articular features include subcutaneous nodules which occur over bony prominences, and inflammation[3] of the sclera (scleritis), arteries (arteritis), and pericardium (pericarditis), and pleural effusion.

The disease can have wide-ranging effects on the family, relationships may change and sexual problems may develop between loved ones. Rheumatoid arthritis requires a multidisciplinary approach from doctors, nurses, physiotherapists, occupational therapists, orthotics and prosthetic departments, social workers and orthopaedic surgeons, to enable patients to remain as independent as possible. Disease management aims to suppress the inflammatory process, relieve pain, promote optimum function and to reduce the psychological and social consequences (Hill 1995).

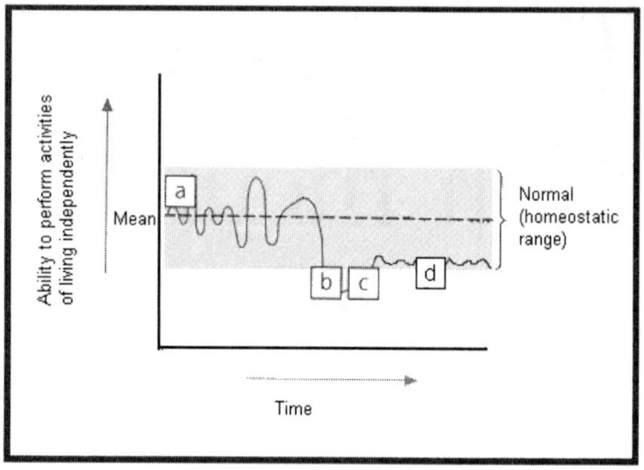

Care

Treatment is provided by the multidisciplinary team. The physician's primary role is to diagnose and prescribe treatment. Referrals are then made to other members of the team.

Mary had both knees aspirated and injected with a long-acting corticosteroid. X-rays of her hands, feet and chest were taken. Methotrexate 7.5 mgs once weekly was prescribed. This is a second line drug which is used to slow down the progression of rheumatoid arthritis and help prevent joint damage.

Mary was referred to the rheumatology nurse practitioner for education about the disease and its treatments. Explanation of the benefits, possible side-effects and the monitoring process of methotrexate were given,

[2] Refer to pp.469–70 for a description of synovial joints.
[3] Refer to p.80 and Table 4.1 for details of inflammation.

Figure 10.1 Rheumatoid arthritis. (a) Physical independence. The ability of the self to perform the activities of living. (b) Imbalance. In rheumatoid arthritis this is caused by flare, or slowly progressing disease, with particular consequences for mobility. (c) Effects of therapies – medical/surgical/nursing/physiotherapy/occupational therapy/self-help – to improve mobility and to facilitate the meeting of needs. (d) Successful return to physical independence. There is, however, no cure for rheumatoid arthritis and a partial restoration of mobility may only be achievable in many people as the disease progresses.

together with a shared care booklet. Referrals were also made to the physiotherapist and the occupational therapist.

The physiotherapist aims to maintain physical function and to teach the patient how to exercise their joints. Treatments such as heat, cold and hydrotherapy are carried out in the physiotherapy department.

The occupational therapist assesses the activities of daily living and teaches joint protection and energy conservation. Aids and appliances can be provided to make carrying out the activities of daily living easier.

The role of the nurse is to provide education about the disease and to allow patients and family the opportunity to discuss issues with the nurse; to monitor some of the 'disease-modifying drugs' and to provide access to specialist knowledge and advice through a telephone helpline; to act as a liaison between the community, hospital and the patient; to assess the patient, identify their needs and to coordinate their care. All these roles will overlap to enable the patient to manage their arthritis with support from the team (see Figure 10.1).

The rationale for investigations made, and drug treatments, are as follows:

INVESTIGATIONS

Rheumatoid factor: this is an antibody.[4] If it is present in the serum, patients are termed seropositive and if not, seronegative. A high titre of this antibody may suggest severe disease if combined with clinical features of inflammation and other high risk factors of developing erosions (Brennan *et al.* 1996).

ERYTHROCYTE SEDIMENTATION RATE (ESR)

This is often raised in the acute phase and is a non-specific indicator of an inflammatory response. It is used to assess disease activity in conjunction with the clinical features of active inflammation e.g. pain, stiffness and swelling.

X-RAYS

These are used in the diagnosis of rheumatoid arthritis, and to monitor the severity and assess the progression of the disease. Changes are seen early in the disease in the hands and feet as erosions usually appear first in the feet. It is not uncommon for X-rays to be normal early in the disease.

Drug treatment

Analgesics are used to help control the pain.

Non-steroidal anti-inflammatory drugs are used to help relieve the pain, stiffness and inflammation but have no effect on the disease progression.

[4] Refer to p.272–3 for a definition and description of antibodies.

Second-line drugs are used as 'disease-modifying agents' with the aim of inducing and maintaining remission. They are potentially toxic, so patients need to be carefully monitored. Regular blood and urine tests, together with direct questioning of the patient, should be carried out to detect possible side-effects.

Corticosteroids are used in low doses, together with a second-line agent in some patients. Side-effects are common, specifically with high-dosage and long-term use.

Intra-articular steroid injections are very effective for flares and where only one or two joints are troublesome. They can significantly reduce pain and swelling in a joint for several months.

Further information

Rheumatoid arthritis is an auto-immune disease where the body's own natural defence system starts to recognize the body's own connective tissue as foreign and attacks it instead of protecting it. It affects more females than males (3:1) The cause is not known and it is thought to be due to a combination of genetic and environmental factors. These may include specific trigger events such as trauma, stress, infection and immunization. A patient may have a predisposition to developing rheumatoid arthritis but it is not directly inherited.

Rheumatoid arthritis is much more common in young women than young men under the age of 40 and this suggests the possible role of the female sex hormones. Remission often occurs in pregnancy but it has a tendency to flare shortly after birth.

The Arthritis and Rheumatism Council literature suggests that 30 per cent of people who develop rheumatoid arthritis will recover completely within one to two years. Sixty per cent will continue to have problems intermittently with slowly progressing disease and between 5 and 10 per cent will become severely disabled.

Main reference

Clancy, J. and McVicar, A. J. 1995 *Physiology and anatomy: a homeostatic approach*. London, Edward Arnold.

Other references

Brennan, P., Harrison, B., Barret, E., Chakreavarty, K., Scott, D. Silman, A. and Symmons, D. 1996 A simple algorithm to predict the development of radiological erosions in patients with early rheumatoid arthritis: prospective cohort study. *British Medical Journal* 313, 471–6.

Hill, J 1995 Patient education in rheumatic disease. *Nursing Standard* 9 (25), 25–8.

The case of a man with chronic obstructive airway disease: chronic bronchitis

11

Helen Thomas

Learning objectives

1 To explain the normal control of breathing and homeostatic alterations in chronic obstructive airways disease

2 To demonstrate an understanding of the care required by someone hospitalized because of acute exacerbation of chronic bronchitis

3 To explain the rationale for care required by someone with an acute exacerbation of chronic bronchitis

---

Case presentation

George is 72 years old and suffers from chronic obstructive airways disease because of chronic bronchitis. He lives alone in a one-bedroom, ground-floor flat. George has smoked all his life until two years ago when he finally gave it up. George takes regular inhaled bronchodilators and has a peak flow meter which he rarely uses. His mobility is moderately limited because of shortness of breath but he still manages to live relatively independently.

Over the past few days George has experienced increased shortness of breath at rest, increased cough with the production of yellow purulent sputum, and has become bedridden from generalized malaise. His GP has been called to visit him at home.

On examination the GP notes immediately that George has great difficulty in holding a conversation because of severe shortness of breath. Assessment of vital signs yields the following results:

- Pulse 110 beats/minute (normal = 60–100 beats/minute)
- Blood pressure 170/95 mmHg (normal = 160/90 at age 70)
- Respiratory rate 32 breaths /minute (normal = 12-15)
- Temperature 37.8°C (normal = 36.9–37.2°C)

George also has a continuous cough with purulent sputum expectoration. Exaggerated breathing movements with the use of accessory muscles are apparent and George complains of exhaustion. His best peak expiratory flow recording is well below that predicted for his age and height. He has been unable to eat properly for some days.

The GP decides that George needs hospital admission and calls for an ambulance.

Case background

Chronic obstructive airways disease (COAD) refers to a clinical syndrome which is characterized by obstruction to the flow of air in the distal airways. It does not refer to one uniform disease, but several pathological conditions:

1 Chronic bronchitis
2 Emphysema
3 Chronic asthma

Chronic bronchitis by definition is based on the presence of a productive cough on most days, for at least three months in the year, over a period of at least two years. The term refers simply to the symptoms of cough and sputum. It is an indicator of mucus hypersecretion. Airway obstruction commonly accompanies chronic bronchitis and is responsible for dyspnoea[1], as evidenced by the increased breathing rate and pulse rate.

Cigarette smoking is overwhelmingly the most important factor in the genesis of chronic bronchitis. Tobacco smoke irritates the bronchial tissue. This initiates the inflammatory response[2] bringing about an increase in glandular tissue, oedema and congestion. At a cellular level, smoking causes metaplasia of the bronchial epithelium, loss of cilia, and increased production of mucus (Figure 11.1). Because of the loss of both serous glands and ciliary action, large quantities of viscid secretions are produced that

All referals are from the main reference at the end of this case study.
[1] Refer to p.317 for notes on obstructive airway disease.
[2] Refer to p.80 and pp.267–9 for an overview of the inflammatory response.

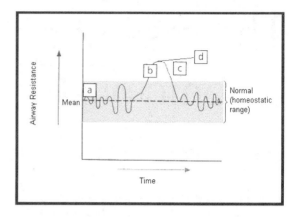

Figure 11.1 Chronic obstructive airway disease. (a) Normal airway resistance. (b) Increased airway resistance caused by excessive mucus production, airway hypertrophy and/or inflammation, etc. (c) Intervention to reduce airway resistance via bronchodilator and expectorant therapy, relieve anxiety, etc. (d) Failure of intervention to correct increased airway resistance leading to a further deterioration of alveolar ventilation and gas exchange.

are difficult to expectorate. The tendency to retention of secretion encourages bacterial growth which provokes further attacks of acute inflammation. Severe chronic exacerbations lead to obliteration of some alveoli and bronchioles. Atmospheric pollution and poor socioeconomic status are also linked to the aetiology of chronic bronchitis.

Care

Before the ambulance arrives the GP will normally administer a bronchodilator via a nebulizer device while reassuring George that his hospital admission is temporary until he is back on his feet again. It will be explained to George that he has developed a severe chest infection and must be treated with intravenous antibiotics. Nebulized bronchodilators directly stimulate beta-2 receptors causing bronchodilation and immediate symptomatic relief from dyspnoea (Figure 11.1).

During transportation to the hospital the paramedic will administer 28 per cent oxygen via a facemask and monitor haemoglobin saturation with oxygen with pulse oximetry (acceptable saturation >90 per cent; saturation is decreased in bronchitis).

The initial aims of treatment are to assess the degree of respiratory failure and correct hypoxia. Assessment tools include pulse oximetry, arterial blood-gas analysis, clinical examination and chest X-ray interpretation. Pulse oximetry provides non-invasive, continuous monitoring of haemoglobin saturation with oxygen.[3] This gives useful information about the adequacy of oxygenation; however, it does not provide information about carbon dioxide retention. It is vital that nurses are able to interpret pulse-oximetry readings in relation to the patient's clinical status. Arterial blood-gas analysis gives specific details of the arterial partial pressures of oxygen and carbon dioxide, pH, and bicarbonate levels which are used to determine the adequacy of alveolar ventilation and acid-base balance. Clinical examination and chest X-ray interpretation indicate the location and distribution of consolidation caused by the infection. The patient also needs to be reassured and supported throughout this acute phase in order to reduce anxiety and, therefore, alleviate cardiovascular changes and reduce respiratory effort exacerbated by increased airway resistance. This can be achieved by the nurse and physician through close supervision and effective communication skills.

Oxygen therapy increases the fraction of inspired oxygen, thereby increasing the alveolar partial pressure of oxygen and correcting hypoxia. In health, the rate and depth of breathing is determined primarily by the arterial partial pressure of carbon dioxide as detected by chemoreceptors. Chronic bronchitis sufferers chronically retain carbon dioxide above normal partial pressures (the elevated values become the normal homeostatic norm) and their respiratory drive is therefore determined primarily

by the decreased arterial partial pressure of oxygen. It is essential that patients suffering from chronic obstructive airways disease are not given more than 28 per cent oxygen. Higher fractions of inspired oxygen can increase the arterial partial pressure of oxygen to levels which remove the drive for breathing, leading to hypoventilation or apnoea.[4] Therefore, the administration of higher fractions of oxygen must be very carefully considered.

On admission to hospital, George receives the following treatment:

- Administration of 28 per cent oxygen via a facemask and continuous pulse oximetry
- Nursed in an upright comfortable position
- Sputum collection pots are made available and George is actively encouraged to expectorate. A specimen of sputum is sent for microbiology, sensitivity and culture
- Administration of a regular nebulized bronchodilator with pre- and post-peak flow recordings
- Venous cannulation and administration of a regular course of intravenous antibiotics
- Chest X-ray
- Arterial blood-gas analysis, full blood count
- Referral for dietary assessment

Nursing the patient in an upright position facilitates the expansion of the chest, minimizing respiratory effort, and helps alleviate sensations of dyspnoea. It is the role of the nurse and the physiotherapist to encourage effective sputum expectoration. This is essential to clear airway obstruction, minimize hypoxia, and prevent further infection. Sputum culture identifies the specific invading bacterial organism and confirms sensitivity to antibiotic therapy. Prior to culture results which may take several days, most infections are assumed to be *Haemophilus influenzae* and are treated as such to instigate antibiotic therapy at the earliest opportunity. The administration of antibiotics via the intravenous

---

[3] Refer to pp.307–9 for a discussion of the carriage of oxygen from the blood.

[4] Refer to pp.311–14 for notes on the control of breathing.

route is indicated in patients with severe respiratory distress since it provides rapid therapeutic plasma levels in a controlled environment.

A comprehensive dietary assessment is required on admission to evaluate the nutritional status of the patient. An inadequate dietary intake prior to admission may increase the susceptibility to infection and decrease the ability to fight infection by suppressing the function of the immune system. Dietary advice is required to prescribe adequate nutrient intake while the patient is hospitalized and advice for home provision ready for discharge. All patients suffering from chronic respiratory disease should have their inhaler technique checked regularly and be given information about the status of their condition. Long-term effective management relies heavily on appropriate self-management which can only be achieved through education and efficient communication.

Further information

In the UK, chronic bronchitis occurs in approximately 15 per cent of men and 5 per cent of women. About 30,000 people die of COAD each year and 30 million working days are lost per annum.

'Blue bloaters' and 'pink puffers': traditionally, chronic bronchitis sufferers with hypersecretion associated with inflammatory damage to airways and impaired respiratory drive became known as 'blue bloaters'. Another group of patients with airway obstruction caused predominantly by emphysema retained their respiratory drive and were known as 'pink puffers'. This distinction no longer exists as it has no pathological validity. Post-mortem and epidemiological studies show that the two groups have similar changes in their lungs. However, the terms 'blue bloater' and 'pink puffer' may reflect variations in ventilatory control, and therefore these groups may require different clinical management.

Main reference

Clancy, J. and McVicar, A. J. 1995 *Physiology and anatomy: a homeostatic approach*. London, Edward Arnold.

Further reading

Clancy, J. and McVicar, A. J. 1997 Hypoxia: a respiratory imbalance. *British Journal of Theatre Nursing* 6 (16) 15–20.

The case of a young woman with secondary hypertension

12

Carolyn Galpin

> *Learning objectives*
>
> **1** To revise the role of the kidney in maintaining blood pressure
>
> **2** To identify the consequences of secondary hypertension
>
> **3** To understand the rationale for the nursing care and management of a patient with hypertension

Case presentation

Mrs Martin, aged 28 years, has been to her doctor with symptoms of tiredness, anorexia, nausea and headaches. Her blood pressure was found to be 160/110 mmHg (normal 120/80 mmHg) and a blood sample showed raised urea concentration (normal 3.0–7.0 mmol/L), raised creatinine concentration (normal 40–110 umol/L), and lowered haemoglobin concentration. After tests she was subsequently diagnosed with renal vascular hypertension caused by renal artery stenosis.

Case background

Stenosis can be caused by fibromuscular hyperplasia of the arterial wall leading to narrowing of the lumen of the arteries. Initially stenosis of the renal arteries causes a decrease in blood pressure within the kidneys and a reduced blood flow distal to the narrowing. The homeostatic response by the kidneys to the hypotension is to release renin from the juxtaglomerular cells in the walls of the afferent arteriole:

- Macula densa cells in the walls of the distal convoluted tubule (DCT)[1] of the nephron monitor sodium levels in the filtrate. When the sodium level falls, because of the reduced glomerular filtration rate (GFR), the macula densa cells stimulate the juxtaglomerular cells (against which they lie) to secrete the enzyme renin. Stretch receptors in the afferent arteriole also participate in the regulation of renin secretion.[2]

All referals are from the main reference at the end of this case study.
[1] Refer to pp.330–1 for a description of the distal nephron.
[2] Refer to pp.238–40 for a discussion of the role of renin in blood-pressure regulation.

- Renin acts on the plasma peptide, angiotensinogen, and converts it to angiotensin I. Angiotensin I is further converted by the angiotensin-converting enzyme (ACE) in the pulmonary endothelial surfaces to angiotensin II. Angiotensin II is a powerful vasoconstrictor.
- Angiotensin II also stimulates the adrenal cortex to release aldosterone, a hormone that causes the reabsorption of sodium, and consequently water, by the DCT.[3]
- This adaptive response is made to maintain adequate renal perfusion needed for vital kidney function. However, the response was triggered by renal hypotension not by systemic hypotension. These adaptive responses, therefore, cause systemic arterial blood pressure to increase. If hypoperfusion of the kidney continues, the renin-angiotensin system continues to cause vasoconstriction and salt and water reabsorption, thus maintaining hypertension and promoting fluid overload (Nowack and Handford 1994).

Diminished glomerular filtration causes a disturbance in body-fluid composition and symptoms such as nausea and anorexia. This is readily detected as being of renal origin by monitoring plasma creatinine and urea concentrations, as excretion of these substances is particularly dependent on the presence of adequate glomerular filtration[4] (see Figure 12.1). Anaemia arises when inadequate renal perfusion prevents the normal production of erythropoietin, the hormone responsible for promoting red blood-cell production.[5]

Care

Rapid pressure increases cause damage to small arterioles and can lead to hyperplastic

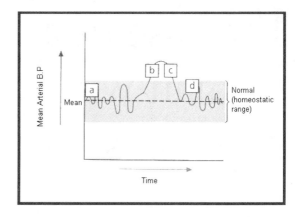

Figure 12.1 Secondary hypertension. (a) Mean blood pressure in a healthy person, fluctuating within its homeostatic range (appropriate for age). (b) Secondary hypertension: blood pressure is increased as a result of the activation of the renin-angiotensin system arising from poor renal blood perfusion. (c) Decreased blood pressure caused by the prevention of the generation of the vasoconstrictor angiotensin II, using an angiotensin-converting enzyme inhibitor (ACE inhibitor: e.g. captopril). (d) Blood pressure restored to normal via good pharmacological control.

arteriosclerosis. The complications are severe and include cardiac infarction, or hypertrophy, and/or failure; cerebral thrombosis and/or haemorrhage; nephrosclerosis and renal failure.

The main treatment in this case is pharmacological to control aggressively the hypertension and to slow renal deterioration. The use of ACE inhibitors may be particularly useful for their effect on intraglomerular pressure (Brundage 1992). Other anti-hypertensives such as calcium channel-blockers, diuretics, or centrally acting agents may also be useful. Artery stenosis can be corrected surgically although renal damage may have already occurred by the time the diagnosis is made.

A further aim is to control the risk factors such as dietary salt intake. Patients who are overweight are put on to low calorie diets, and smokers are advised to stop or at least cut back on smoking.

---

[3] Refer to pp.444–5 for notes on the action of aldosterone and to Figure 15.15 for a diagrammatic summary of the role of renin and aldosterone in blood-pressure regulation.
[4] Refer to p.334 for a discussion of creatinine clearance.
[5] Refer to pp.176–7 for a description of the process of erythropoiesis.

People with kidney disease must also restrict the amount of potassium and protein in their diets to lower the potential of retaining urea (from protein metabolism) and increasing K^+ concentration. They must also restrict their sodium and fluid intake to prevent fluid overload.

Further information

Renal artery stenosis is a condition found more frequently in young women; it rarely occurs in men (Muir 1988). The complications are severe and include end stage renal disease.

Main reference

Clancy, J. and McVicar, A. J. 1995 *Physiology and anatomy: a homeostatic approach.* London, Edward Arnold.

Other references

Brundage, D. 1992 *Renal disorders.* St Louis, Mosby Year Book.

Muir, B. L. 1988 *Pathophysiology: an introduction to the mechanisms of disease,* 2nd edn. New York, John Wiley.

Nowack, T. J. and Handford, A. G. 1994 *Essentials of pathophysiology.* Oxford, W.C.Brown.

13 The case of a man with acute renal failure

Carolyn Galpin

Learning objectives

1 To revise the general principles of the regulation of body fluids

2 To revise the role of the kidneys in fluid and electrolyte balance

3 To identify the consequences of electrolyte imbalance and fluid imbalance

4 To recognize the clinical signs of acute renal failure (ARF)

5 To understand the rationale for the nursing care and management of a patient with ARF

Case presentation

Mr Turner, aged 56 years, had major abdominal surgery two days ago and is recovering apparently uneventfully in a surgical ward. The nurse monitors his vital signs and his fluid intake and output. He has an intravenous infusion of dextrose/saline running at 100ml/hour. His blood pressure has stabilized at 115/70mmHg, after having had a prolonged hypotensive episode during the night which was believed to be due to a post-operative fluid-volume deficit. His pulse is 90 (normal 60–90 beats per minute); respirations are 22 (normal 12–16 respirations per minute).

On checking his urinary catheter drainage bag, the nurse observes that only 60ml of urine has been passed during the last four hours (normal urine output for an adult is at least 30ml/hour). There are no obvious signs of blockage or kinking in the catheter, and Mr Turner does not have a distended bladder or feel any bladder discomfort.

Case background

Acute renal failure is a condition characterized by a sudden, rapid deterioration in renal function. There is oliguria (less than 30ml urine/hour or less than 400 ml in 24 hours), or anuria (rarely) (0–100ml urine/24 hours), together with uraemia: raised blood urea levels (McLaren 1996).

The hypotensive episode experienced by Mr Turner reduced both renal perfusion and the glomerular hydrostatic pressure necessary for an adequate glomerular filtration rate (GFR) (normal GFR = approx. 120ml/min).[1] When the filtering mechanism is substantially impaired the kidneys are unable (a) to excrete nitrogenous waste products e.g. urea, uric acid sufficiently, and (b) water, electrolyte, and acid-base balance may also be impaired. The

All referals are from the main reference at the end of this case study.
[1] Refer to p.325, p.328 and pp.334–5 for a discussion of glomerular filtration and its measurement.

failure to excrete excess urea, fluid and electrolytes produces serious and potentially life-threatening problems.[2]

- Fluid overload causes a fluid shift within the body fluid compartments, precipitating pulmonary oedema with dyspnoea and peripheral oedema.[3]
- Expansion of the extracellular fluid compartment (ECF) is exacerbated because of a raised sodium concentration. Consequently there is movement of water from the intracellular fluid across cell membranes into the interstitial spaces.[3]
- Failure to maintain electrolyte balance may lead to raised serum potassium (K^+) levels (hyperkalaemia) (normal K^+ levels = 3.5–5.0 mmol/L) which can cause fatal cardiac arrhythmias (see the case of a man with hyperkalaemia), and is often the cause of death in ARF.[4]
- Failure to excrete the hydrogen ions of 'fixed' acids may result fatally in metabolic acidosis (normal blood pH 7.35–7.45).
- Urea concentrations begin to rise as acute renal failure continues (normal concentrations 3.0–7.0 mmol/L).

A complexity of problems and symptoms arise including anorexia, nausea, vomiting, drowsiness and confusion, alteration in blood-clotting mechanisms, and a compromised immune system. Coma and death will ensue if some form of renal replacement therapy is not instituted e.g. haemofiltration, haemodialysis, or peritoneal dialysis.

Care

The specific goal of care is to manage, as far as possible, the fluid and electrolyte balance, to

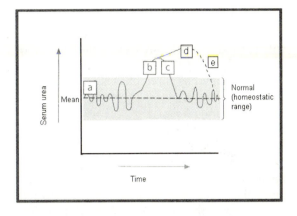

Figure 13.1 Acute renal failure. (a) Normal serum concentration of urea in health arising from normal urea clearance values. (b) Elevated serum urea concentration caused by decreased urea clearance arising from reduced renal filtration in acute renal failure. (c) Return to normal concentration as spontaneous recovery of renal filtration (and urea clearance) occurs. (d) Persistent elevated serum urea concentration caused by a failure of spontaneous renal recovery as a result of acute failure. (e) Intervention (e.g. haemodialysis, etc.) to restore normal urea concentration (plus the normalization of other blood solutes).

restore the acid-base balance,[5] and reduce the rising urea levels until normal renal function returns (see Figure 13.1). Fluid intake is reduced, usually to 500 mls/24 hours plus the previous day's output volume. This is to replace insensible loss and allow for fluid gain caused by oxidative metabolism.[6] Dietary sodium intake is restricted to reduce a positive sodium balance, and thus prevent further fluid shift within the body-fluid compartments. Dietary potassium intake is restricted and ion exchange resins, such as calcium resonium, can be administered to promote K^+ secretion into the colon. High calorie diet provision may also minimize K^+ ions produced by tissue breakdown (Groer and Shekleton 1989). The restriction of dietary protein is necessary to help

[2] Refer to p.344 for further details of homeostatic disturbances in renal failure.
[3] Refer to pp.92–4 for a discussion of the movement of water between fluid compartments.
[4] Refer to p.95 and p.228 for an explanation of the effects of potassium ions on cell membranes and cardiac functions respectively.

[5] Refer to pp.342–3 for a discussion of the role of the kidneys in acid-base balance.
[6] Refer to pp.96–7 for a discussion of water balance.

reduce the production of urea and other nitrogenous products.[7] If necessary, dialysis will be used to artificially restore blood composition to the normal homeostatic status.

ARF usually has three phases:

1 the oliguric phase
2 the diuretic phase
3 the recovery phase.

Renal replacement therapy (RRT) is necessary until the daily serum urea and electrolyte levels reflect the recovery of renal function. This can still be required even when the diuretic phase begins (the oliguric phase can last up to six weeks) because the kidneys are initially unable to concentrate the urine. Strict monitoring is therefore essential throughout the course of ARF. Daily weighing will indicate the amount of fluid retained and which will require removal by ultrafiltration during dialysis, or by haemofiltration.

Nursing care also includes several weeks of psychological support, optimum nutrition to maintain an anabolic state (high calorie plus necessary dietary restrictions, as mentioned) often as total parenteral nutrition, blood transfusions to treat anaemia, the prevention of infection caused by compromised immune system, and the treatment of the consequences of uraemia.

Further information

Whatever the cause, acute renal failure is a life-threatening condition and there is still quite a high mortality rate despite new and sophisticated technological treatment. However, approximately 60 per cent of patients regain most or all of their renal function. Where possible, haemofiltration is the treatment of choice because it can be continuous and so maintain the patient haemodynamically in a more stable state. It can also be used to remove calculated amounts of fluid.

The causes of the persistent reduction in renal function in ARF remain unknown.

Main reference

Clancy, J. and McVicar, A. J. 1995 *Physiology and anatomy: a homeostatic approach*. London, Edward Arnold.

Other references

Groer, M. W. and Shekleton, M. E. 1989 *Basic pathophysiology: a holistic approach*. Baltimore, C.V. Mosby.

McLaren, S. and Watson, R. (eds) 1996 Renal function, cited in Hinchliff, S. M. *Physiology for nursing practice* 2nd. edn. London, Baillière Tindall.

[7] Refer to p.153 for an outline of the metabolism of proteins by the liver.

The case of a girl with acute anxiety: hyperventilation

14

Helen Thomas

Learning objectives

1 To identify the subjectivity of psychological and physiological responses to stress

2 To explain the aetiology of hyperventilation associated with acute anxiety states

3 To demonstrate an understanding of the care required by someone admitted to an accident and emergency department with acute anxiety and hyperventilation

4 To explain the rationale for care of a person with acute anxiety and hyperventilation

Case presentation

Clare is a 19-year-old. She was admitted to the accident and emergency department in a state of severe anxiety. She had been out for the evening with some friends who had brought her into hospital because she was in a state of panic. On admission, Clare was breathing extremely rapidly and her heart rate was raised:

- Pulse 140 beats/minute (normal = 60–100 beats/minute)
- Respiratory rate 42 breaths /minute (normal = 12–18 breaths/minute)

Clare complained of chest tightness, dizziness, and pins and needles in her fingers. She appeared restless and was shaking.

Case background

Hyperventilation is commonly associated with severe anxiety states and in particular panic attacks. Extreme emotional outbursts may be associated with fear or an inability to cope with a particular stressful situation. Responses to such situations are highly individual and may manifest themselves as both psychological and physiological disturbances (refer to the

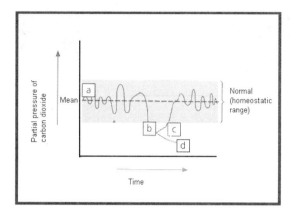

Figure 14.1 Hyperventilation. (a) Homeostatic range of alveolar and arterial carbon dioxide. (b) Decreased arterial partial pressure of carbon dioxide caused by excessive excretion i.e. hyperventilation. (c) Restoration of arterial partial pressure of carbon dioxide to normal range by decreased excretion, promoted by re-breathing from a paper bag. (d) Failure of intervention leading to an excessive decrease in partial pressure of carbon dioxide, as in prolonged hyperventilation leading to respiratory alkalosis.

'Further information' section of this case study).[1]

Acute hyperventilation may present clinically in a variety of forms. While some individuals experience sudden tightness in the throat, gasping for breath, rapid shallow breathing, chest pain, dizziness, and pins and needles, others may encounter far more insidious symptoms such as constant yawning or sighing, and excessive sniffing which may lead to 'chronic hyperventilation' with time.

Severe anxiety states can stimulate higher centres in the brain via the stimulation of the sympathetic nervous system, causing an increase in the rate and depth of breathing. The release of endogenous adrenaline and noradrenaline prolong this action which serves to prepare the body for 'flight and fight' (Cannon 1935). This short-term homeostatic control serves to enable us to cope with or adapt to the perceived stressor.

All referals are from the main reference at the end of this case study.
[1] Refer to Chapter 22 for a discussion on the subjectivity of stress.

Hyperventilation reduces the alveolar partial pressure of carbon dioxide and, conversely, raises the alveolar partial pressure of oxygen (Figure 14.1). Arterial partial pressures reflect these changes, although hyperventilation has little impact on the amount of oxygen carried by the blood since haemoglobin is normally saturated anyway.[2] The decreased arterial partial pressure of carbon dioxide observed with prolonged hyperventilation leads to respiratory alkalosis, and also cerebral anoxia (the cerebral circulation responds to reduced carbon dioxide content by vasoconstriction). Fainting re-establishes control of the respiratory rate and depth which both decrease (hypoventilation) until homeostatic balance is restored.

Care

Clare was asked to breathe in and out of a paper bag. This continued for ten minutes until Clare stated that she no longer felt dizzy and her respiratory rate and depth had dropped to normal parameters. Throughout this time the nurse remained with Clare, offering her support and reassurance. In order to slow down the rate of Clare's breathing the nurse encouraged her to relax her shoulders, breathe from the diaphragm, and concentrate on breathing out.

Once Clare had established a normal breathing pattern and appeared settled the nurse tried to establish the cause of her anxiety attack. Clare explained that she had begun to panic while having a drink in a very busy bar. She had experienced feelings of severe claustrophobia and an inability to breathe.

As Clare had not experienced these feelings of acute anxiety before she was not referred to a psychiatrist. Her GP was notified of her admission for follow-up at a later stage.

Hyperventilation causes excessive amounts of carbon dioxide to be blown off by the lungs. This excessive excretion of carbon dioxide

[2] Refer to pp.306–9 for revision of gas carriage by blood and pp.316–17 for notes on hypo/hyperventilation states.

causes a respiratory alkalosis. Under normal circumstances the partial pressure of carbon dioxide in arterial blood is the major factor controlling the rate and depth of respiration. In contrast, moderate fluctuations in oxygen levels have little effect on the rate and depth of breathing. Clare was asked to breathe in and out of a paper bag so that she would re-breathe her own expired air which would contain more carbon dioxide than atmosphere air (Figure 14.1). This in turn would increase the levels of carbon dioxide in her blood with only a slight reduction in oxygen levels. Once Clare had calmed down, her respiratory rate and depth decreased until the normal partial pressures of carbon dioxide in her bloodstream were re-established.

People who exhibit signs of acute anxiety or panic may harm themselves or others and need to be supervised closely. Constant support and reassurance is required to alleviate the distress experienced by the individual. If possible, a quiet environment and a calm approach help to diffuse the intense feelings of fear and 'loss of control'.

Psychiatric referral is not usually required when there is no previous history of acute anxiety states, although each case must be assessed individually. GP follow-up can be useful for evaluating the individual's recovery from the attack and for discussing the general levels of stress in the patient's life and their coping strategies. Advice can also be given on what to do in the event of raised anxiety levels or another attack.

Further information

Stress is a multidimensional concept consisting of both psychological and physiological components. The unique physical and psychological signs and symptoms that can be observed in individuals suggest that there is a strong subjective element involved in the response to stress. The complex nature of stress has incited conflicting definitions from the scientific disciplines which emphasize (with varying degrees of significance) the impact of the environment as a stressor, social stressors, and the individual character of responses.[3] The perception of stressors and the subsequent psychophysiological response to stress varies greatly between individuals and is unique to each person. While some individuals, like Clare, may experience the traumatic symptoms of hyperventilation during severe anxiety states, other individuals may not.

Main reference

Clancy, J. and McVicar, A. J. 1995 *Physiology and anatomy: a homeostatic approach*. London, Edward Arnold.

Other reference

Cannon, W. B. 1935 Stresses and strains of homeostasis. *American Journal of Medical Sciences*, 189 (1).

[3] Refer to pp. 637–9 for definitions of stress.

15

The case of a man with benign prostatic hyperplasia

Louise Fuller

Learning objectives

1 To define the condition benign prostatic hyperplasia

2 To identify the signs and symptoms of benign prostatic hyperplasia

3 To understand the reasons for treatment of the acute chronic obstruction of urine

Case presentation

Bill is now 72. He retired from the civil service seven years ago. He had enjoyed good health but had suffered from nocturia (passing urine two to three times each night) for several years, but had not consulted with his GP, believing it to be a normal phenomenon one suffers as one ages.

Eight months ago, Bill had suffered a bout of urinary retention. He had a hot bath which resolved it. The next morning he consulted his GP. On examination nothing abnormal was detected in the urine or kidneys. There was no bladder distension, but the GP found a large soft prostate on rectal examination. As Bill had suffered an episode of retention, the GP decided to refer him to a consultant. In the meantime a blood specimen was obtained to check urea, electrolytes and prostatic acid phosphatase (PSA) concentrations. Urea and electrolytes were normal but the prostatic acid phosphatase was raised. This enzyme is a marker for prostatic tumour and its elevated blood concentration is indicative of hyperplasia.

A biopsy, under general anaesthetic, found no evidence of malignancy and Bill was diagnosed as suffering from benign nodular hyperplasia of the prostate. Because of his symptoms he was placed on the waiting list for a transurethal resection of prostate (TURP).

While waiting for his operation, Bill suffered a series of urinary tract infections which were treated with the appropriate antibiotics. At one stage Bill had to be catheterized as he had another episode of urinary retention.

Bill eventually had his TURP but his problems did not end. He suffered from further episodes of urinary tract infections with frequency of micturation and increasing nocturia. A rectal examination by the GP confirmed a still sizeable prostate remnant. He was therefore referred back to the consultant. In the meantime Bill lost weight and generally felt unwell.

A cystoscopy carried out under general anaesthesia showed a grossly trabeculated bladder and a residual volume of urine of 950 mls. A further TURP was then performed. This was successful. Bill has had no problems since and has gained weight.

Case background

As middle age approaches, the prostate gland gradually hypertrophies (see Figure 15.1) and because of its anatomical position surrounding the urethra,[1] the enlargement can obstruct urine flow, resulting in homeostatic imbalances of the bladder, urethras and kidneys.

The resulting symptoms from this obstruction can be many, depending on whether the obstruction is chronic or acute. In chronic obstruction, frequency, dribbling, urgency, nocturia, incontinence, hesitancy and stream can be experienced. In acute obstruction anuria, pain and a distended palpable bladder usually present as an emergency episode and the man needs to be catheterized.

The excretion of waste products is essential for the homeostatic maintenance of all systems in the body.[2] If urine is not excreted sufficiently, dilatation of the collecting system proximal to the blockage will occur.[3] The urethra becomes distorted and displaced, so urine flow is impeded. If the bladder wall is affected, dependent areas occur, so there is incomplete emptying with urinary stasis. This predisposes to the formation of calculi and infection, as stagnant urine serves as a culture for bacterial growth. The urine pressure and/or infection can backtrack up the urethra to the kidneys with the result of dilatation of the pelvis and calyces,

All referals are from the main reference at the end of this case study.
[1] Refer to Figure 18.5 for an illustration of the prostate gland.
[2] Refer to pp.321–3 for an introduction to excretion in maintaining cellular homeostasis.
[3] Refer to pp.328–33 for a description of renal anatomy.

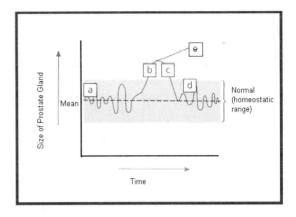

Figure 15.1 Prostatic hypertrophy. (a) Prostate gland size appropriate to the age of the individual. (b) Prostatic hypertrophy. This typically occurs in late adulthood, for as yet unidentified reasons. If excessive, the hypertrophy will induce urethral obstruction and require intervention. Hypertrophy may be indicative of prostatic cancer. (c) Correction of prostatic hypertrophy by surgery and pharmacological therapy. (d) Restoration of prostate size to within normal limits. (e) Persistent prostatic hypertrophy leading to a risk of renal failure.

leading to renal failure if left untreated. If renal failure was to occur, gross disturbance of body fluid composition would arise which would be fatal if uncorrected.[4]

Care

The treatment of benign prostatic hypertrophy is therefore aimed at restoring the normal flow of urine, for example, by catheterization and surgery, and by restoring normal fluid and electrolyte balance, which may be disrupted by the obstructive process.

Bill's first bout of urinary retention, before the prostate hypertrophy was diagnosed, was reversed by a hot bath. The hot bath probably caused muscle relaxation. But this would not resolve the retention, subsequently experienced with more advanced hyperplasia.

[4] Refer to pp.345–6 for a discussion of the consequences of renal failure.

Further information

The reason for hypertrophy of the prostate is not clear but it may be related to changes in oestrogen and androgen levels which occur as men age.

The problems caused by benign prostatic hypertrophy can be relieved by medication and/or surgery. The exact treatment depends on the severity of the symptoms. The medication acts either by relaxing the muscles surrounding the prostate, bladder and urethra or by gradually reducing the size of the prostate and so widening the urethra and allowing urine to flow more freely.

Selective alpha blockers[5] (for example, prazosin tamsulosin) may also be used to block the actions of the sympathetic activity and relax the smooth muscle surrounding the prostate and urethra.

Sex-steroid activities may also be treated pharmacologically. For example, the five alpha reductase inhibitor, finasteride, blocks the formation of dihydrotestosterone and induces the shrinkage of hyperplastic tissue in the prostate.

Main reference

Clancy, J. and McVicar, A. J. 1995 *Physiology and anatomy: a homeostatic approach*. London, Edward Arnold.

[5] Refer to pp.332–33 for a discussion of sympathetic activity in micturition.

PART TWO

Child case studies

The case of a neonate with fetal alcohol syndrome

16

Sue Sides

Learning objectives

1 To revise normal fetal development

2 To understand the influence of the uterine environment on fetal development

3 To understand the rationale for care of the neonate with fetal alcohol syndrome

4 To increase awareness of the need for health promotion in relation to pregnancy

---

Case presentation

Thomas is one day old. He was born in hospital following a normal delivery, to his mother Mary who is 32 years old. Mary is a known alcoholic with a history of hospital admissions related to alcohol abuse. Mary's antenatal care was managed by her GP although she could only be persuaded to visit him twice during her pregnancy. Home visits were often unsuccessful and Mary attended the antenatal clinic only once. She refused to allow the community midwife to visit her at home and she also refuses to disclose the whereabouts of her family or the identity of the father of her baby.

Thomas was of below average birth weight, i.e. <2.5kg. He has the characteristic appearance of a baby born to a mother with a history of alcohol abuse: a small head (microcephaly), and an abnormal facial appearance – a short nose with a flat bridge and a small jaw with thin lips.

Thomas is an 'irritable' baby and although born at term, he has a poor sucking reflex. Thomas has been diagnosed with fetal alcohol syndrome (FAS).

Case background

FAS results from the teratogenic effects of alcohol on the developing embryo.[1]

All referals are from the main reference at the end of this case study.
[1] Refer to pp.588–90 for a discussion and identification of common teratogens.

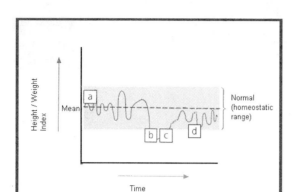

Figure 16.1 Fetal alcohol syndrome. (a) Height–weight index within homeostatic range: development within normal centile range for gestational age. (b) Homeostatic imbalance: height–weight index indicates development below normal centile range. (c) Nutritional intervention to restore homeostatic balance within altered parameters of FAS. (d) Failure of growth rate to 'catch up'. Height–weight index remains low for age.

Alcohol may be considered as one of the most important preventable causes of fetal damage. Teratogenic levels of alcohol have not yet been established but it is thought that all amounts are potentially harmful and damage is likely to be dose-related. Alcohol may have the following fetal effects:

- Acute fetal poisoning which may lead to spontaneous abortion
- Damage to the developing nervous system resulting in learning disabilities which may vary in severity from mild communication problems to severe mental retardation
- Low birth weight with failure to 'catch up' in spite of subsequent adequate nutrition (see Figure 16.1)
- Abnormal development of facial features – thought to be caused by a defect of midline mesodermal cells in early embryogenesis[2]
- Cardiac malformations
- Limited joint movement and other skeletal problems

[2] Refer to pp.565–7 for a description of embryo development.

Care

Thomas will be hypersensitive to sensory stimuli because of the neurotoxic effects of alcohol and will therefore require nursing intervention to be kept to a minimum until his tolerance improves.

Thomas will be nursed in a quiet nursery environment, where the frequency of nursing intervention will be minimized and tailored to his tolerance pattern. Nursing contact will be initiated slowly by tactile stimulation.

FAS infants often assume overextended postures, particularly of the limbs, and therefore swaddling in a soft warm blanket may be helpful.

Pharmacological intervention may be required if Thomas exhibits signs of withdrawal from alcohol: increasing irritability, fever, vomiting/diarrhoea, a high pitched cry, sweating and seizures. Chloropromazine may be used to decrease irritability but care must be taken to avoid oversedation and depression of respirations, heart rate[3] and sucking rate.

Adequate nutrition is essential. Breast-feeding is encouraged where the mother has stopped drinking, otherwise small regular bottle feeds will be given. Mary will be encouraged to feed Thomas herself if at all possible.[4] If sucking is poor, nasogastric feeding may be necessary.

Daily weight and head circumference will be recorded to establish growth.

Thomas will be carefully monitored for signs of infection. This is essential as FAS infants are more susceptible to infection as a result of poor uterine nourishment, exposure to maternal pathogens and the immaturity of the immune system.

Wherever possible Thomas's mother should be actively involved in his care to facilitate maternal attachment. She will need much support as it is likely that, in the initial stages at least, Thomas's response to his mother, or indeed to any caregiver, will be negative and

[3] Refer to p.377 for notes on the role of the brain stem in regulating vital signs.
[4] Refer to p.112 for a discussion of the additional demands of breast-feeding on energy intake by the mother.

potentially distressing. Failure to establish an early infant–mother relationship may have serious consequences for their later relationship.

Mary will require on-going support and education if she is to be able to care for her son. Thomas is Mary's first child, and as she has not attended any parent craft classes it is highly likely that she will require help with these skills. The relationship that is established with health professionals in the neonatal period will have a marked impact on her parenting skills and the willingness and ability to alter her lifestyle.

Discharge from hospital will need to be planned carefully with Mary's support and cooperation to facilitate close liaison between the multidisciplinary team and ensure a safe and supportive home environment. The services of the community midwife, health visitor, social services, GP and the voluntary services will be required.

Thomas will be reviewed at regular intervals in the paediatric outpatient department.

At an appropriate time, information about self-help groups, such as Alcoholics Anonymous, will be given to Mary.

Further information

The causative factor in FAS seems to be the direct toxic effect of ethanol or its metabolite acetaldehyde.

The effects of FAS are permanent. Little is known about the long-term risks associated with fetal exposure to alcohol. Preschool children with FAS are often described as being hyperactive and learning disability is not uncommon, although this may not become apparent until the primary school stage. Most children have a friendly personality and some older ones have been noted to have abnormal minor motor movement. Hyperactivity may persist.

The most important effect of FAS is central nervous system dysfunction and there appears to be a correlation between observable abnormalities (already described) and the degree of cognitive impairment.

Interestingly, studies indicate that although the IQ of children born to alcoholic mothers is often within normal limits, these children persistently demonstrate academic failure, illustrating the influence of biopsychosocial interactions.

In some cases FAS may be diagnosed retrospectively when a baby presents with 'failure to thrive' (FTT).

Main reference

Clancy, J. and McVicar, A. J. 1995 *Physiology and anatomy: a homeostatic approach*. London, Edward Arnold.

17 The case of a child with sickle cell disease

Theresa Atherton

> ## Learning objectives
>
> **1** To understand the genetic basis for the inheritance of sickle cell trait and sickle cell disease
>
> **2** To understand the role of sickle haemoglobin as an evolutionary homeostatic mechanism for reducing the morbidity and mortality associated with malaria
>
> **3** To explore the altered physiological processes involved during an acute painful crisis
>
> **4** To understand the rationale for care in relation to disordered physiology

Case presentation

Conrad is a six-year-old boy with chronic sickle cell disease, admitted by his GP, with an acute painful crisis affecting the joints and soft tissues of his hands and feet (see case background). He is accompanied by his parents and ten-year-old brother Leroy.

On admission, Conrad is screaming and clings to his mother when anyone approaches. He appears to be in great pain and is unwilling to talk. His father states that his son woke up that morning complaining of a sore throat and earache. He went back to bed and slept for the morning but has now started vomiting. His parents have tried to encourage Conrad to drink but with little success.

Examination proves difficult because of Conrad's distress. Observations that are obtained show an axilla temperature of 38.5°C and a pulse of 145 bpm. His hands and feet appear swollen and warm to the touch.

Conrad's parents both have sickle cell trait; their other son, Leroy, has neither the trait nor the disease.

A more detailed assessment later reveals that Conrad has had six previous crises since diagnosis. He has otherwise been in good health although he has never achieved urinary continence.

Case background

Two alleles (genes) are required to determine most characteristics of an individual.[1] The alleles are located on each member of a chromosome pair. For the healthy individual

All referals are from the main reference at the end of this case study.
[1] Refer to pp.598–9 for a discussion of alleles.

with a homozygous (i.e. both alleles identical) genotype of HbA/HbA (HbA = adult haemoglobin), all haemoglobin will be of the normal adult form.[2] For the individual with a heterozygous (i.e. the alleles are different) genotype of HbA/HbS (HbS = sickle haemoglobin), erythrocytes will have approximately 60 per cent of HbA and 40 per cent of HbS (Franklin 1990). This is because HbA and HbS are codominant: that is, both are expressed. Such individuals are said to be carriers of the sickle cell trait.[3] Individuals with the homozygous genotype HbS/HbS will have sickle cell disease (SCD), in which all haemoglobin is of the sickle cell type. It is of interest to note that the only difference between HbA and HbS is the substitution of just one amino acid in the protein component.[4]

In Conrad's case, he has inherited both sickle cell alleles from his parents and therefore has sickle cell disease. Leroy, on the other hand, has been fortunate in not inheriting the abnormal alleles from his parents and is therefore healthy (homozygous HbA/HbA).

Children born with SCD are initially healthy because of the presence of fetal haemoglobin (HbF) which has a greater affinity for oxygen, thereby facilitating normal function.[5] Symptoms can commence after three months of age although more usually between six and nine months, as by this time HbF has been replaced by HbS and passive immunity, acquired from the mother begins to wane prior to sensitization of the child's own immune system.

The primary problem with SCD is that under conditions of reduced oxygen tension (hypoxia), lactic acid accumulates. This leads to metabolic acidosis.[6] Under such conditions, the

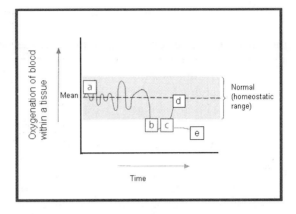

Figure 17.1 Sickle cell disease. (a) Oxygenation of the tissue within homeostatic norms, appropriate to the activity of the tissue. (b) Tissue hypoxia. Generally, this can arise through poor oxygenation of blood in the lung, by anaemia or by a failure to deliver sufficient blood to the tissue (ischaemia). In sickle cell disease, obstruction of the microcirculation prevents adequate blood flow. (c) Hypoxia is reversed by the use of oxygen therapy and rehydration or by pharmacological intervention to improve blood supply. (d) Restoration of tissue oxygenation. (e) In sickle cell disease, the occlusion of vessels by 'clumped' red cells will normally make such interventions only partially effective. Ischaemic pain may persist as oxygenation remains lower than that required for normal tissue function.

relatively insoluble HbS molecules collapse, sticking together to form crystalloid polymers. This alteration of molecular structure causes the characteristic sickle-shaped cells. Sickling leads to a sluggish circulation with clumping and thrombosis occurring in small blood vessels. Obstruction of these vessels leads to ischaemia and necrosis (see Figure 17.1), the effects of which would be related to the site of occlusion. All tissues are at risk[7] and in all cases severe pain is a feature. Such episodes are referred to as 'crises', the frequency of which is variable: up to three a year are the most common. These crises may last from a few days to a few weeks.

The earliest signs of SCD are effects on the bones and soft tissues of the hands and feet. In England 1 per cent of children with SCD

[2] Refer to pp.306–7 for a description of adult haemoglobin.
[3] Refer to p.601 for an explanation of genetic 'carriers'.
[4] Refer to pp.29–31, p.603 and Figure 20.7 for details of DNA structure, and altered protein in sickle haemoglobin respectively.
[5] Refer to pp.307–9 for a discussion of oxygen carriage by blood.
[6] Refer to p.60 for an outline of anaerobic metabolism.

[7] Refer to p.603 and Figure 20.7 for an outline of the systemic effects of sickle cell disease.

develop this complication (Franklin 1990). Conrad's swollen hands and feet would be indicative of occlusion to the blood vessels supplying his extremities. Adequate microcirculation is essential for tissue-fluid formation and reabsorption; interference to this process by a compromised circulation would therefore lead to the accumulation of tissue fluid manifested by swelling and warmth. Ischaemia of tissues would account for Conrad's severe pain.

Sickling is also influenced by the viscosity of blood. Hypertonicity of plasma, as in dehydration, increases the concentration of HbS within the erythrocytes and thereby increases the likelihood of sickling occurring. Conrad has had a reduced fluid intake for the preceding 24 hours, the effects of which have been accelerated by vomiting. It is quite likely that he has become dehydrated. Conrad's pyrexia (elevated body temperature), resulting in increased metabolic demands, will have further depleted his fluid reserves.

Sickling in the renal medulla eventually causes ischaemic damage to the nephron, reducing its ability to concentrate glomerular filtrate. This would be indicated by abnormal renal function tests. This may often lead to nocturnal enuresis (wetting the bed at night), or difficulty in the child gaining urinary continence, as in Conrad's case.

Care

There is currently no cure for sickle cell disease: management is symptomatic and preventive. Sickle cell crisis can be reversed under conditions of adequate oxygenation and hydration.

Conrad would be placed on bed rest to minimize his energy expenditure thereby reducing oxygen demands; although appropriate rest will only be achieved through effective analgesia. Pain management is particularly difficult during sickle cell crisis. It is usual for opiates to be prescribed, in particular pethidine, because of its ability to provide the rapid

relief of severe pain.[8] Patient controlled analgesia (PCA) may be considered, as this has been successfully used with children during acute painful crisis, (Grundy et al. 1993). Non-invasive techniques such as distraction therapy, relaxation techniques and/or guided imagery should be considered as complementary to pharmacological methods of pain control. Age-appropriate assessment would be used to evaluate the effectiveness of any intervention.

While on restricted activity, the passive movement of Conrad's limbs will be needed to improve microcirculation to his extremities in order to improve tissue-fluid reabsorption and so reduce oedema. Ambient temperature would be controlled to help Conrad keep comfortably warm; extra clothing or bedding will be provided as required. It must not be forgotten that Conrad is pyrexial because of his infection and increased metabolic demands. The comfort value achieved by a warm environment should not compromise attempts to reduce his body temperature. The regular observation of axilla temperature, pulse and respirations would be commenced to monitor his progress throughout the period of crisis; the frequency of these is reduced once Conrad's condition stabilizes.

Blood investigations would be carried out to establish baseline values for comparison of such parameters as haemoglobin, white cells, platelets, urea, electrolytes and liver-function tests. Blood would also be required for grouping[9] and saving, should transfusion become necessary.

Conrad has been complaining of a sore throat and earache, and it is therefore quite likely that he has developed an upper respiratory tract infection. Screening will be carried out which might include a midstream/clean-catch specimen of urine, blood cultures (for bacterial and viral studies), throat swab, and nasal swab/sputum specimen. Antibiotics will then be administered to treat any primary infections or to prevent secondary infection.

---

[8] Refer to pp.620–2 for an overview of opioid therapy.
[9] Refer to pp.185–7 for discussion of blood groups.

Dehydration would be assessed through serum electrolyte analysis. In Conrad's case, vomiting would preclude rehydration with oral fluids, therefore an intravenous infusion of dextrose saline (with added potassium if required) would be commenced at the earliest opportunity. The monitoring of fluid input and output is vital during the rehydration process, particularly as Conrad's ability to concentrate urine is impaired.

Erythrocyte parameters are usually unchanged in vaso-occlusive attacks in comparison to those seen in steady state levels. Unless Conrad was thought to be acutely hypoxic, such as in heart failure, oxygen therapy would not be of any great benefit. The blood supply to the affected area is compromised because of sickling, therefore oxygen would not be able to reach that area to reduce tissue hypoxia.

Transfusion/exchange transfusion may be considered should Conrad's crisis not respond to conventional treatment.

Further information

Sickle cell disease has its origins in those parts of the world where malaria is endemic, i.e. Africa, the Caribbean, countries bordering the Mediterranean and the Asian subcontinent. It is thought that the genetic mutation of the haemoglobin gene which produces sickle cell trait may have persisted as an evolutionary tactic to help overcome the malarial parasite.

Following a bite from an infected mosquito the malarial parasite enters the blood stream. It gains entry to red blood cells and survives by feeding on protein derived from haemoglobin (Hb). As parasites multiply the cell eventually bursts, releasing large numbers of parasites into the blood stream. These newly released parasites repeat the cycle.

The presence of HbS slows down the rate of multiplication of the malarial parasite by collapsing readily when the cell becomes invaded. Thence the carriers of sickle cell trait (HbS/HbA) are protected from the potentially serious complications associated with sickle cell disease by the presence of normal adult haemoglobin. This is known as the 'heterozygote advantage'.

Ideally, all infants would be tested for sickle cell at birth, but this may not be cost-effective in populations where there are relatively small numbers of individuals from ethnic minorities. As a compromise, it is usual for screening to be carried out only on children in a high-risk category, i.e. those from ethnic minority groups with a high incidence of SCD or other haemoglobinopathies. It is accepted practice for all patients from 'at risk' groups to be routinely tested for SCD if they require an anaesthetic as part of their treatment.

The time of diagnosis is crucial so that preventive measures can be introduced that will help to reduce morbidity and mortality, particularly in the first year of life. Following diagnosis of SCD, the family will be instructed to prevent the child from becoming chilled, to provide extra fluids during hot weather or when the child undertakes excessive exercise, to make sure the child rests at the first sign of any pain and to drink plenty of fluids (Baughan 1985). The child would be placed under the care of a paediatrician, who would monitor the steady state pattern and ensure early hospitalization should it prove necessary.

There are two other forms of crisis that can occur with SCD: anaemic and sequestration crises (Franklin 1990). Anaemic crises are uncommon but are seen more often in children than in adults. Infection causes the bone marrow to stop making red cells, causing the signs and symptoms of severe anaemia. Sequestration crises are rare but are also precipitated by infection. Their occurrence is almost exclusively in children. There is blood engorgement of the spleen, which may eventually fill the whole abdomen. This dramatically reduces circulating blood volume, causing symptoms of acute shock. Treatment is the administration of packed red blood cells and the possible removal of the spleen to prevent future attacks.

Main reference

Clancy, J, and McVicar, A. J. 1995 *Physiology and anatomy: a homeostatic approach*. London, Edward Arnold.

Other references

Baughan, A. Hughes, A. Paterson, K. and Stirling, L. 1985 *Manual of haematology*. Edinburgh, Churchill Livingstone.

Franklin, I. 1990 *Sickle cell disease: a guide for patients, carers and health workers*. London, Faber & Faber.

Grundy, R., Howard, R. and Evans, J. 1993 Practical management of pain in sickling disorders. *Archives Diseases of Childhood* 69(2), 256–9.

The case of a baby boy with necrotizing enterocolitis

18

Stevie Boyd*

Learning objectives

1 To revise the possible complications associated with preterm birth

2 To understand the pathophysiology of necrotizing enterocolitis

3 To be aware of the different treatments for each stage of the disease

Case presentation

Jamie was born at 28 weeks gestation, weighing 1.24 kg to a primiparous mother. The pregnancy had been reasonably uncomplicated until a small antepartum haemorrhage progressed to preterm labour.

Drugs to inhibit labour were contraindicated as it was necessary to deliver the baby in optimal condition before maternal blood loss had a detrimental effect on the fetus.

Jamie was in excellent condition at birth but was electively intubated and maintained on minimal ventilatory support for 12 hours.

At 24 hours of age, he developed a pyrexia. His abdomen was becoming distended and an X-ray revealed pneumoperitoneum (free gas under the diaphragm) and pneumotosis intestinalis (intramural bubbles). Intramural gas is mainly hydrogen produced as a product of carbohydrate metabolism by anaerobic bacteria in the gut. Bowel sounds were absent.

A few hours later, Jamie was in obvious pain. This was caused by cellular damage releasing intestinal pain-producing substances,[1] and the distended abdomen was splinting his diaphragm, causing respiratory distress with deteriorating blood gases (hypercapnia, hypoxia and acidosis).

In spite of the absence of bloody stools and only minimal bilious aspirate, a diagnosis of perforated necrotizing enterocolitis was made.

Case background

Jamie's pyrexia was a physiological response to the presence of pyrogens released by pathogens;[2] his neutrophil count was raised, indicating bacterial invasion.[3]

All referals are from the main reference at the end of this case study.
*This case study is adapted from Boyd, S.C. (1997) Necrotizing enterocolitis: a biological science orientated case study. Journal of Neonatal Nursing 3: 16–21.

[1] Refer to p.6I5 for notes on pain-producing substances.
[2] Refer to p.518 for notes on pyrogens producing fever.
[3] Refer to pp.179–80 for a description of neutrophilia on a bacterial invasion.

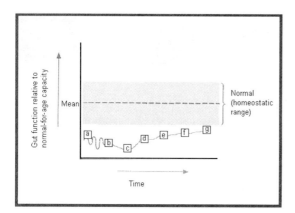

Figure 18.1 Necrotizing enterocolitis. (a) Compromised gut function caused by prematurity. (b) Ischaemic episodes soon after birth. (c) Necrotizing enterocolitis, with infection. (d) Clinical stabilization using antibiotics, respiratory support and analgesia/sedation. (e) Resection of gut and formation of stoma. Total parenteral nutrition. (f) Intestinal continuity restored. Oral feeding. (g) Gut function *almost* fully recovered with normal-for-age tissue functions.

The pathogenesis of necrotizing enterocolitis (NEC) is multifactorial, but it is an acquired neonatal disorder following a combination of vascular, mucosal and toxic insults to an immature gut. It involves ischaemic gut mucosa followed by infection with gas-forming organisms. It has been suggested (Grosfeld *et al.* 1991) that toxic free radicals[4] of oxygen could be the cause of cell damage (Figure 18.1).

Enteral feeds have been implicated (Glomella 1992: p.372) in the pathogenesis because of the osmolarity and lack of immuno-protective factors in formula milk.

The incidence is 0.3–15 per 1000 live births and is more common in boys. The majority weigh less than 1500 grams at birth.

NEC has become a major problem in the last 25 years, with higher survival rates because of improved clinical expertise producing an ever-enlarging group of increasingly premature and low birth-weight infants. It is the most serious surgical disorder among infants who have endured intensive care intervention, and is a significant cause of morbidity and mortality.

Care

There are five stages of NEC: these range from mild abdominal distension without systemic symptoms to a fulminant course of sepsis caused by endotoxin release, disseminated intravascular coagulation, collapse and death (see Table 18.1). Similarly, management may entail simple medical intervention or it may require aggressive multisystem medical support as well as surgery.

Jamie had the severest form: advanced necrotizing enterocolitis. He had deteriorating vital signs and electrolyte imbalance with shock syndrome. There was evidence of perforation and generalized peritonitis and pneumoperitoneum. He required clinical stabilization followed by surgical intervention to prevent rapid deterioration and possible death.

When Jamie first became ill, he was prescribed triple antibiotics: broad spectrum penicillin and gentamicin, plus metronidazole which has specific activity against anaerobic bacteria found in the gut.

Hydration and nutrition were achieved parenterally, and a nasogastric tube on free drainage was *in situ* to decompress the bowel.

To facilitate adequate gaseous exchange and to prevent further deterioration of his blood gases, Jamie was reintubated and ventilated, and a morphine infusion to provide analgesia and sedation was commenced.

Jamie required urgent surgery. The diagnosis was confirmed when the ileum was found to be perforated and meconium was found in the peritoneum. Three centimetres of gut were removed, and a dysfunctioning double-barrelled ileostomy was formed to allow the bowel to heal.

Post-operatively, Jamie was ventilated for 48 hours and antibiotics were continued for ten days. Analgesia was achieved with bupivicaine

---

[4] Refer to p.585 for an explanation of the 'free radicals' theory of ageing.

Table 18.1 The stages of NEC and specific treatment

Stage	Signs and symptoms	Treatment
I (Pre-NEC)	Non specific; apnoea, bradycardia, lethargy, unstable temperature, gastric aspirate, occult blood in stools, non-specific ileus	Nil by mouth, continuous nasogastric drainage, 2–4 hourly observations of vital signs, removal of umbilical catheters, antibiotics, monitor fluid balance, septic screen and blood picture, serial X-rays. If cultures negative after 3 days, stop antibiotics and recommence feeds.
IIA (mild NEC)	As above, plus prominent abdominal distension +/- tenderness, absent bowel sounds, gross blood in stools, ileus and dilated loops of bowel with focal areas of pneumotosis intestinalis on X-ray	As above, but continue antibiotics for 10 days, give TPN and withhold enteral feeds for at least 7–10 days after X-ray shows clearing of pneumotosis, supplemental oxygen if hypoxic, platelet transfusion if thrombocytopaenic, treat acidosis, ? surgical opinion
IIB (moderate NEC)	Mild acidosis and thrombocytopaenia, abdominal wall tenderness +/- palpable mass, extensive pneumotosis and early ascites and intrahepatic venous gas	As above
IIIA (advanced NEC)	Respiratory and metabolic acidosis requiring ventilation for apnoea, hypotension and decreased urine output, spreading oedema, erythema and in duration of abdomen, prominent ascites, no perforation	As above and respiratory support as dictated by blood-gas levels and clinical status, blood pressure support/volume expansion – inotropes, colloid to replace ongoing fluid losses, blood and platelet transfusions
IIIB (advanced NEC)	Deteriorating vital signs and lab indices, shock syndrome and electrolyte imbalance, evidence of perforation with peritonitis and pneumoperitoneum	As above, then surgery when clinically as stable as possible. Peritoneal drainage can be done on ward or laparotomy with resection +/- stoma or anastomosis

via an epidural catheter – causing less respiratory depression than systemic opioids.

Total parenteral nutrition (TPN) via a percutaneous central line was begun after two days to deliver substrates directly into the circulation, promoting anabolism and providing for normal growth while allowing the gut to heal. Enteral feeds to supplement TPN were introduced cautiously at two weeks once the bowel had recovered and gastric aspirates were clear (Figure 18.1).

Stoma care, overseen by a nurse specialist, was meticulous; it was crucial to preserve the integrity of Jamie's preterm skin.

The parents required empathetic emotional and psychological support and needed every encouragement to be involved in Jamie's care.

Jamie's recovery was long and stormy with several relapses and complications requiring five months of hospitalization. He was readmitted a few weeks after discharge for emergency surgery as his stoma had prolapsed. His intestinal continuity has now been restored. He appears to be well and thriving; hopefully he will not be one of the significant number of neonatal unit graduates who suffer psychomotor delay requiring remedial education.

Further information

Necrotizing enterocolitis is potentially life-threatening and may have long-term consequences. Efforts to understand the aetiology and possible preventive measures are a priority. Epidemiological studies revealed that most associated factors describe events in a population of high-risk neonates, and that no maternal or neonatal factors, other than prematurity, exist when there is an inverse relationship between the risk of NEC and gestational age.

Prevention is best achieved by managing high-risk conditions in pregnancy and preventing hypoxic premature births. If prematurity cannot be avoided, maternal steroids may be of benefit to accentuate lung maturation by stimulating surfactant synthesis. Intrinsic surfactant production helps to prevent respiratory distress syndrome: the most common cause of neonatal mortality in the UK. Antibiotics and giving only breast milk do not prevent the disease; but the slow introduction of feeds, minimizing episodes of hypotension, hypoxia and hypothermia and the use of umbilical catheters appear to have some relevance.

Recent studies have shown that the administration of oral IgA, to exert an immunoprotective effect on the gastrointestinal tract, may decrease the incidence.

Jamie's NEC was most probably the result of his premature birth, although it was unusual for it to develop so early and without enteral feeding. He did have an umbilical catheter *in situ* for a very short time, but was not exposed to any other risk factor. No organisms were found in his blood culture or surface swabs.

Main reference

Clancy, J, and McVicar, A. J. 1995 *Physiology and anatomy: a homeostatic approach*. London, Edward Arnold.

Other references

Glomella, T. L. 1992 Neonatology. *Necrotizing Enterocolitis*, Chapter 66. Connecticut, Appleton & Lange.

Grosfeld J. L., Cheu, H., Sch-later, M., West R. W. and Rescorla, F. J. 1991 Changing trends in necrotizing enterocolitis, *Annals of Surgery* September, 214 (3), 300–6.

The case of a child with vitamin D deficiency-induced rickets

19

Theresa Atherton

Learning objectives

1 To examine the role of vitamin D in relation to calcium homeostasis

2 To explore the physiological effects of calcium deficit in rickets

3 To understand the rationale for care in relation to disordered physiology

4 To explore the significance of health promotion in the management of rickets

Case presentation

Hussein is 13 months old, the second child of Ali and Sue. The family live with paternal grandparents in a three-bedroom semi-detached house on a large private housing estate. Following his birth, Hussein progressed satisfactorily and was well cared for by his extended family; but the health visitor noted that on every occasion she visited, Hussein was inside the house, wrapped up snugly in his pram, even during the summer months. His mother seemed unperturbed by this, stating that Hussein needed to keep warm: she was frightened that he might catch cold if she left him outside in the garden too long.

Hussein was breast fed until he was four months old when the health visitor suggested that they might begin weaning. As the family was strictly vegan, the health visitor suggested that Hussein receive vitamin supplements to prevent the possibility of vitamin and mineral deficiency;[1] both parents agreed.

When Hussein was seven months old, the health visitor became concerned about him as his previously steady weight gain began to level out at the third centile, although he appeared happy and healthy. The GP was informed and examined Hussein, finding nothing organically wrong with him. He suggested an assessment of Hussein's nutritional intake to ensure an adequate diet and continuation of the vitamin supplements.

Hussein is now 13 months old. On a routine visit, the health visitor finds him up and toddling about. She notes that he appears somewhat unsteady on his feet and seems to have 'bandy' legs, more prominently than might have been expected for a child of his age.

All referals are from the main reference at the end of this case study.
[1] Refer to pp.108–10 for a discussion of vitamins and minerals.

On close questioning, Hussein's mother admits that she has not been giving her son the vitamin supplements as he didn't seem to like them. She also states that he is reluctant to take solid food, preferring to feed from the breast. The health visitor examines Hussein and detects that his skull bones seem unusually soft and his anterior fontanelle shows no sign of closure. She mentions her concerns to the GP, who subsequently sees Hussein and refers him to the local paediatrician who confirms a diagnosis of vitamin D deficiency-induced rickets.

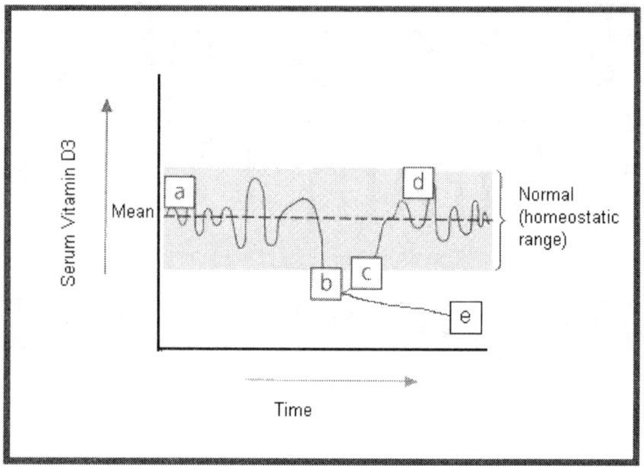

Case background

Primary or 'exogenous' rickets results from a combination of a nutritional deficiency of vitamin D and a lack of exposure of the skin to sunlight (see Figure 19.1). The clinical features of the disease relate directly to the effects that vitamin depletion has on calcium homeostasis[2], particularly the effects on the musculoskeletal system of the growing child.

There are a number of vitamin D derivatives. The most common and most important is cholecalciferol (vitamin D_3). This is predominantly manufactured by ultraviolet irradiation of 7-dehydrocholesterol in the skin through exposure to sunlight but it is also naturally present in most dairy products, fish-liver oils and eggs (some milk and food products are also fortified with extra vitamins).

Hussein's parents are vegan's and have therefore excluded all animal products, including dairy produce, from their son's diet.[3] Hussein's reluctance to take weaning foods and his parent's reliance on breast-feeding may have unwittingly precipitated a vitamin D deficiency. Their failure to administer the vitamin supplements has compounded the problem. Hussein's dietary deficiency may also have been complicated by his mother's

Figure 19.1 Rickets. (a) Vitamin D_3 availability optimal for normal bone mineralization. (b) Insufficient vitamin D_3 arising from either inadequate dietary intake or insufficient synthesis by the skin because of lack of exposure to UV radiation in daylight. Bone mineralization will be reduced, increasing the risk of rickets in children. (c) Increased availability of vitamin D_3 as a result of dietary change, vitamin supplements and/or increased exposure to direct sunlight. Vitamin availability should be restored to homeostatic range (d). (e) Failure to correct vitamin D availability increases the risk of a persistent 'bowing' of the limb bones in older children.

reluctance to let her son be exposed to natural daylight because of her fear that he may become unwell.

The mineral calcium is essential for the normal development of the musculoskeletal system and the dentition of children from the early stages of fetal growth to the end of puberty. The maturation of the skeletal system of the child involves the gradual deposition of inorganic salts within cartilaginous bone, providing strength and durability. If salts and, in particular, calcium fail to be deposited in the matrix of epiphyseal cartilage (growth area of bones) these cells hypertrophy (abnormally enlarge) and pile up irregularly to many times their normal thickness. The affected area becomes weakened and is easily compressed, deformed or displaced. X-rays of Hussein's spine show early signs of lateral curvature (scoliosis) which would be compatible with the weakening of his vertebral column as he began to become weight-bearing.

[2] Refer to pp.341–2 for details of calcium homeostasis.
[3] Refer to pp.111–17 for a discussion of nutritional requirements.

The increased secretion of parathyroid hormone[4] in response to lowered serum-calcium levels found in rickets results in excessive bone demineralization, causing the characteristic 'bowing' of the shafts of long bones. In Hussein's case his rickets has not been diagnosed until after he has become fully weight-bearing, which has resulted in the 'bandiness' observed on physical examination. If severe rickets occurs before a child begins to walk, it produces a combination of bowed thighs and knock-knees.

Although the bowing of the upper limbs is rare in rickets, X-rays may reveal wrist changes compatible with the infant putting pressure on the limbs while crawling and attempting to push themselves into the standing position. Although Hussein is not currently exhibiting signs of bowing in his upper limbs, X-rays of his wrists revealed characteristic changes, indicating that rickets was active during the first few months of his life. Such changes frequently precede 'bowing' of the lower limbs, but the disease may continue to remain undetected until the more dramatic characteristics of the latter are observed.

The skull is sometimes soft to the touch, the bone recoiling when the pressure is released (craniotabes). This effect can be likened to the feel of a table-tennis ball when the surface is gently compressed and then released. This might account for the sensation of 'sponginess' that the health visitor noted on examining Hussein. This phenomenon can be considered to be normal at the suture lines in a young infant, but may be indicative of rickets when it occurs away from the suture lines. Closure of the fontanelles[5] is usually delayed which would account for the fact that Hussein's anterior fontanelle can still be clearly felt.

The management of Hussein's rickets will be vitamin supplements with regular monitoring of his disease state until healing has taken place. This may be for up to two years following the commencement of treatment. The therapeutic dose of vitamin D for the management of Hussein's rickets would be in the region of 10–40 micrograms daily (McClaren 1992).

The dietician would assess Hussein's nutritional status and calculate his developmental requirements, taking into account the dietary restrictions imposed by a vegan diet. Providing an adequate diet would not be difficult but would require careful planning. Following the resolution of his rickets, Hussein's parents may be advised to continue vitamin supplements until growth was complete, as strict vegetarianism may predispose to other vitamin deficiencies or a recurrence of rickets during periods of rapid growth. It is difficult to establish an estimated average requirement (EAR) of vitamin D as most of it is produced in the skin even in places with poor sunlight, but it is thought to be in the region of 10 micrograms daily for most age groups, assuming the child is exposed to sunlight (McClaren 1992).

The amount of vitamin D in breast milk is 0.8 micrograms per 100 mls in comparison to 0.15 micrograms in cow's milk.[6] As can be seen, breast-feeding would not be contraindicated as long as appropriate solid foods were introduced into Hussein's diet to supplement the vitamin D intake.

Hussein's parents would be counselled about their practice of swaddling. Although culturally acceptable, the importance of exposing their son's skin to sunlight would need to be emphasized, as this is the primary site of vitamin D production.

Blood tests may be undertaken intermittently as a means of assessing Hussein's nutritional status, particularly with regards to vitamin D, as metabolite levels can indicate the progress of the disease. This is particularly evident with the metabolite alkaline phosphatase which is elevated early in the deficiency state as a result of bone demineralization. Hypocalcaemia appears late in the

[4] Refer to pp.441–2 for a description of parathyroid hormone.
[5] Refer to p.578 for details of the infant skull.

[6] Refer to p.115 and to Table 6.8 for an analysis of breast milk.

disease process, but monitoring levels may be useful during the recovery process.

Hussein may have X-rays to monitor bone-healing, but these would be potentially unhelpful in the early stages as any detectable changes appear late in the disease process.

Further information

Vitamin D (calciferol) is commonly regarded as a hormone rather than a vitamin because of the similarity of its chemical structure to members of the steroid family and its manufacture and mode of function within the body.

The deficiency of vitamin D is called rickets during childhood and osteomalacia in the adult; its clinical features will vary accordingly. Although rickets is more commonly considered to be a primary disease induced through a nutritional deficiency of vitamin D, there are other forms of endogenous deficiency such as those sometimes found in malabsorption syndromes, liver disease, chronic renal failure and some genetic disorders.

Rickets is a health problem readily influenced by environmental factors affecting both the physical and psychological development of the child. Children living in densely populated areas, in basements or in high-rise flats, may unwittingly have minimal exposure to natural daylight. This situation is potentially compounded by a preoccupation with sedentary indoor-play activities. It is important that such children are encouraged to play outside when opportunities arise.

Nutritional disturbances of vitamin and mineral intakes are largely preventable: primary prevention is the establishment of homeostatic balance. The most significant impact of such nutritional disturbance on children is during the first few months and years of life.

Dietary-induced rickets was once rare when milk was fortified with vitamin D, but today, factors such as the cessation of welfare vitamin D supplements, prolonged breast-feeding with the delayed introduction of weaning, an adherence to a strict vegetarian diet that is low in vitamin D, and marginal exposure to sunlight through excessive swaddling, may all contribute to the presentation of dietary-induced rickets.

Main reference

Clancy, J. and McVicar, A. J. 1995 *Physiology and anatomy: a homeostatic approach*. London, Edward Arnold.

Other reference

McClaren, D. 1992 *A colour atlas and text of diet-related disorders*, 2nd edn. London, Mosby.

The case of a child with hypertrophic pyloric stenosis

20

Theresa Atherton

Learning objectives

1 To revise the applied developmental anatomy and physiology of gastric digestion

2 To explore the impact of disordered physiology on homeostatic processes

3 To understand the principles of rehydration and surgical intervention as a means of restoring homeostatic function

4 To understand the rationale underpinning pre- and post-operative care

Case presentation

Charlie is seven weeks old, and was born following an uneventful pregnancy. During the past week, he has begun vomiting in between breast feeds but otherwise appeared well. The vomiting has now settled into a regular pattern following most feeds, and although initially ravenous following these episodes, Charlie has become increasingly listless. His mother reports that his nappies have not been as wet as usual and his stools have become more formed.

On examination, Charlie does not object to being handled: he remains quiet and lethargic. His eyeballs appear sunken and his anterior fontanelle is depressed. There is also a loss of skin elasticity and his abdomen appears distended.

Charlie is given a 'test feed' during which his abdomen is palpated. An olive-shaped mass is felt in the epigastric region, and visible peristaltic waves are observed moving across the abdomen from left to right. Following the test feed, Charlie proceeds to have a 'projectile' vomit. His history and current signs and symptoms lead to a diagnosis of hypertrophic pyloric stenosis.

Case background

The pyloric sphincter guards the exit from the stomach. It is in a semi-permanent state of contraction, and consequently allows some gastric fluid to pass rapidly into the intestine. But it prevents solid and semi-solid food leaving until gastric digestion is completed (for further details of the anatomy and physiology of the stomach, see footnote[1]).

Pyloric stenosis is an obstructive disorder caused by hypertrophy (excessive growth) of the circular smooth muscle of the pyloric sphincter. This results in the narrowing of the lumen which partially/completely impedes the passage of chyme from the stomach. Initially the infant appears well but as the condition progresses the classic signs of acute intestinal obstruction become apparent. These are characterized by colicky abdominal pain, nausea, vomiting and abdominal distension.

Initially the pyloric sphincter is able to relax sufficiently to allow the passage of partially digested milk, but delayed emptying of the stomach eventually leads to an increase in gastric pressure. As the capacity of the stomach is exceeded, periodic vomiting ensues.[2] Vomitus may be foul-smelling, containing undigested milk, mucus and curds of stale insoluble milk protein. Small flecks of fresh and digested blood (which have the appearance of coffee-grounds) may be seen in the later stages which are as result of chronic inflammation of the gastric mucosa. The infant will initially be hungry after vomiting and will be eager to repeat the feed, often settling well with no further vomiting.

As hypertrophy of the pylorus continues, obstruction becomes more severe. There is a rapid build-up of accumulated chyme and more frequent episodes of vomiting. Projectile vomiting usually develops as a later sign, vomitus being forcefully ejected 2 to 4 feet from the sitting infant.

During test-feeding, the abdomen is gently palpated to detect the enlarged pyloric sphincter. The term 'pyloric tumour' refers to the hypertrophied pyloric muscle, and does not in any way infer a malignant (cancerous) condition. Therefore caution must be exercised when using such terminology with both Charlie's parents and the general public.

Visible peristaltic waves[3] moving across the abdomen from left to right are sometimes evidence of powerful peristaltic action as the muscular stomach wall attempts to overcome the obstruction. Assessment of pain in infants is difficult, but it is to be expected that Charlie would experience some degree of colicky pain from such exaggerated peristalsis.

Infants have a higher percentage of total body water (approximately 75 per cent), in comparison to older children. Their daily exchange of extracellular fluid is also much greater, ranging from between 25–45 per cent of total body weight per day (Campbell and Glasper 1995). Such physiological differences can be attributed to differences in body and organ size and the immaturity of those physiological processes concerned with fluid balance. This leaves minimal reserves of body fluid when intake is increasingly impaired, such as in Charlie's case. Protracted vomiting soon leads to dehydration.

Homeostatic mechanisms aimed at water conservation will therefore be activated. Charlie's urine output will diminish as antidiuretic hormone, released by the posterior pituitary gland, results in renal salt and water retention.[4] Fluid will also be increasingly absorbed from the colon accounting for the semi-formed stools of a breast-fed infant becoming more solid and less frequent.

If the fluid balance is not maintained, fluid rapidly becomes lost from both extra- and intracellular compartments. Loss of skin elasticity, sunken eyeballs and a depressed fontanelle are all signs of increasing dehydration. Charlie's listlessness could be directly attributable to such dehydration and malnourishment.

It is usual for the stomach to be decompressed following diagnosis. This is achieved by passing a nasogastric tube which may be

All referals are from the main reference at the end of this case study.

[1] Refer to pp.134–8 for details of gastric anatomy and physiology.

[2] Refer to pp.157–8 for a discussion of vomiting response.

[3] Refer to p.134 and Figure 7.8 for a description of peristalsis.

[4] Refer to p.338 for a discussion of the role of ADH in water balance.

aspirated (artificial removal of gastric contents) and/or left on free drainage. This facilitates the removal of residues of partially digested milk curds and helps prevent the accumulation of gases that may be causing Charlie's abdominal distension. Occasionally, normal saline lavages (fluid introduced slowly into the stomach via a nasogastric tube and then aspirated) are prescribed, which fully empty and cleanse the stomach. Normal saline is isotonic with body fluids therefore no water absorption takes place across the gastric mucosa.[5] Charlie will not be allowed oral fluids so arrangements will be made for his mother to express her breast milk.

Pyloromyotomy is the standard surgical technique for the management of hypertrophic pyloric stenosis. It consists of a longitudinal incision being made through the circular muscle fibres of the pylorus down to, but not including, the sub-mucosa. This has the effect of enlarging the lumen of the sphincter, thereby overcoming the intestinal obstruction (see Figure 20.1).

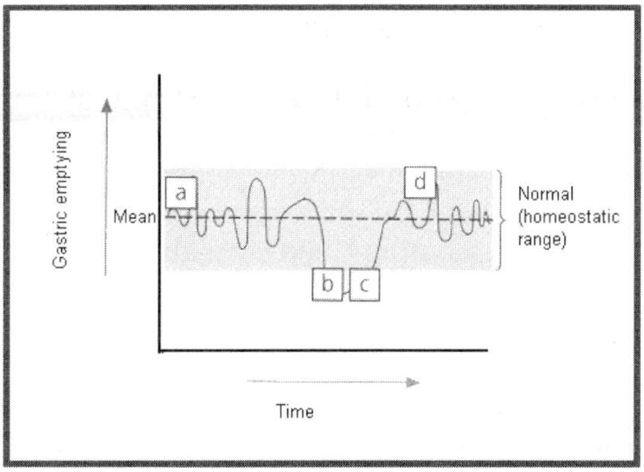

The primary focus of preoperative care is restoring fluid and electrolyte balance and the safe preparation of child and family for surgery. Where there are no obvious signs of dehydration, surgery is performed without delay. In Charlie's case, his fluid balance and general condition were assessed and closely monitored. Surgery was postponed for 24 hours to allow rehydration to take place.

Charlie will be weighed, an assessment made for fluid replacement and an intravenous infusion commenced. Frequent assessment of fluid and serum electrolyte levels will be undertaken during the rehydration process. A glucose and saline solution will initially be administered to correct water depletion and prevent further malnutrition. Blood glucose may drop rapidly as vomiting and imposed starvation deplete stores of glycogen.[6]

Hyponatraemia (low serum sodium) and reduced chloride levels invariably accompany

Figure 20.1 Pyloric stenosis. (a) Gastric emptying appropriate to feeding pattern, and age, of the infants. (b) Inadequate gastric emptying. In this case, this was a result of a hypertrophied pyloric sphincter. (c) Surgical reduction of the hypertrophied sphincter (pyloromytomy). (d) Normal gastric emptying restored (following healing).

water depletion and are to be expected following prolonged periods of vomiting and nasogastric aspiration. Potassium is also lost in the same way as sodium, but replacement of potassium via the infusion will normally be delayed. This is because during dehydration, potassium is withdrawn from dehydrated (hyperosmotic) cells creating a state of hyperkalaemia (high serum potassium). Renal function is restricted during dehydration (as previously described) and, as a result, excretion of this excess potassium is impaired. During rehydration, extracellular potassium moves back into the intracellular compartment. Correction of potassium depletion via the intravenous infusion will therefore only take place once normal renal function has been restored. Care must be taken to ensure that the required infusion rate is adhered to, as too rapid an administration may lead to heart block and cardiac arrest.[7]

Hydrogen ions are also lost from vomit and gastric aspirate (loss of hydrochloric acid). This

---

[5] Refer to p.19 for details of water movement by osmosis.
[6] Refer to p.153 for details of glucose metabolism in the liver.

[7] Refer to p.228, for an analysis of the effects of hyperkalaemia on cardiac cells.

may rapidly lead to metabolic alkalosis. As previously stated, accurate monitoring of electrolyte balance is essential during rehydration to detect imbalances and to allow the prompt adjustment of intravenous therapy.

Vital signs of pulse, respirations, temperature, urine output and weight will be monitored to evaluate the effectiveness of rehydration therapy. Accurate records will be made of all fluid input and output, including any episodes of vomiting and bowel actions. General hygiene for a dehydrated infant is important, particularly of the skin and mouth. Charlie's mouth will be kept moist with boiled water as he will not be taking oral fluids, and his skin will be checked for integrity and signs of pressure. He will be monitored for infection as dehydration and malnutrition may increase his susceptibility. Pain will be assessed using an age-appropriate tool and analgesia will be administered accordingly.

Once Charlie's hydrate ion/electrolytes have been stabilized, he will be prepared for surgery as per local hospital policy.

Postoperatively, normal feeding patterns are usually established within 24 hours, but this may vary according to local practice. Intravenous therapy is discontinued when adequate fluid levels are being taken orally. Using dissolvable sutures negates the need for the removal of stitches. Recovery is usually unremarkable. Charlie will be discharged from hospital on the second to sixth postoperative day, depending on his progress.

Further information

Hypertrophic pyloric stenosis is one of the most common surgical disorders of infancy. It is five times more common in male than female infants, affecting approximately five in every 1000 males and only one in every 1000 females (Cohen 1984). The cause is unknown but there is thought to be some hereditary involvement.

The success rate of pyloromyotomy is high where the correction of fluid and electrolyte imbalance has been achieved preoperatively (Campbell and Glasper 1995). Vomiting may persist in the immediate postoperative period but will gradually diminish as the pylorus adjusts to its normal size and function.

Main reference

Clancy, J, and McVicar, A. J. 1995 *Physiology and anatomy: a homeostatic approach.* London, Edward Arnold.

Other references

Campbell, S. and Glasper, A. (eds) 1995 *Whaley and Wong's children's nursing.* London, Mosby.

Cohen, F. 1984 Clinical genetics in nursing practice. Cited in Campbell, S. and Glasper, A. (eds) 1995 *Whaley and Wong's Children's Nursing.* London, Mosby, page 574.

The case of a boy with lymphoblastic leukaemia

Andrew McVicar and John Clancy

Learning objectives

1 To revise the types of cells found in the blood

2 To recognize the consequences of a failure to maintain control of cell division in the aetiology of leukaemia

3 To understand the basis for chemotherapy

4 To understand the rationale for the care of an individual who is undergoing chemotherapy

Case presentation

Billy is five years old. He had been a healthy 'normal' boy, but has recently been experiencing painful, swollen joints. He has also shown signs of general malaise including pallor, fatigue and fever. Frequent nosebleeds have also proved troublesome, and Billy's mother has noticed small blood spots (petechiae) in his skin. His GP referred him for further examination.

Blood analysis showed that Billy was anaemic and that his platelet count was only 100 000 per mm³ ('normal' is around 250 000 per mm³). The total number of white blood cells was 10 000 mm³, only slightly above normal, but a differential count showed an excess of lymphocytes especially B-lymphocytes.[1] Many of the white cells in the sample appeared immature, or had morphological or chromosomal abnormalities. The subsequent aspiration of a bone-marrow sample showed a preponderance of lymphocytic stem cells (lymphoblasts), and confirmed that Billy had acute lymphoblastic leukaemia.[2]

Case background

Acute lymphoblastic leukaemia is the most common form of leukaemia to occur in children. The cause is unknown and is speculated to involve environmental agents which cause genetic mutation. The typical early onset, the predisposition of children with Down's syndrome to this type of leukaemia, and the high concordance between affected identical twins, suggests that there may also be an inherited component.[3]

In this type of leukaemia there is a change in the behaviour of lymphocyte stem cells within

All referals are from the main reference at the end of this case study.
[1] Refer to p.174 and pp.178–81 for a discussion of white blood cells and differential cell counts.

[2] Refer to pp.192–3 and Table 8.11 for an overview of leucocytes.
[3] Refer to p.605 for a discussion of polygenic inheritance.

the bone marrow. Other cells are crowded out by these rapidly dividing cells, causing the reduced production of red blood cells, platelets and white blood-cell types. The white cells also appear abnormal and immature. Thus:

- decreased red blood-cell synthesis produces anaemia[4] and so pallor and fatigue
- the platelet deficiency reduces the capacity for blood-clotting, leading to episodes of bleeding[5]
- the abnormal white cells are generally more mobile than normal and may infiltrate tissues. Infiltration of the joint capsule, and subsequent inflammation, are responsible for joint-swelling and pain[6]
- the rapid turnover of marrow cells in leukaemia results in the release of excessive amounts of uric acid (a product of nucleic acid catabolism) and the deposition of uric acid crystals in the joints will exacerbate joint pain
- white blood cells, particularly lymphocytes, are central to the body's defence against micro-organisms.[7] The abnormal cells observed in acute lymphoblastic leukaemia mean that Billy will be more susceptible to infection

The main risk to life in leukaemia is internal haemorrhage, especially in the brain, and uncontrolled infection.

Care

Billy was placed on a regime of chemotherapy using anti-leukaemic drugs. Such drugs are very toxic especially at the high doses used. A 'cocktail' of drugs is given. Chemotherapy drugs interfere with cell activities, especially those related to DNA replication and cell

[4] Refer to pp.189–90 for a discussion of anaemias.
[5] Refer to pp.183–5 for a discussion of the clotting process.
[6] Refer to pp.267–9 for a discussion of inflammation.
[7] Refer to pp.274–80 for a discussion of the role of lymphocytes in immune responses.

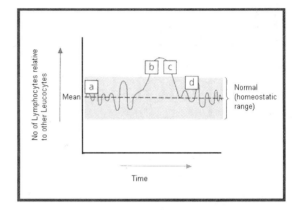

Figure 21.1 Leukaemia. (a) Normal differential cell count; numbers of lymphocytes relative to other white blood cells appropriate for health. (b) Excessive numbers of lymphocytes. Slight elevations could reflect the presence of infection. Pronounced increases, as observed in lymphoblastic leukaemia, result from excessive activity of lymphoblasts in bone marrow. (c) Influence of therapies (chemotherapy and/or radiotherapy) to reduce lymphoblast activity. (d) Successful effects of therapies to restore differential white cell count to normal. The individual will now be in remission.

division, to kill the leukaemic cells. The actions of the drugs are indiscriminate, however, and will also affect healthy cells, so the side-effects are wide-ranging. Observations include watching for signs of neuropathy (e.g. personality change, weakness of handgrip, ptosis of eyelids), of gastro-intestinal disorder (e.g. gastric ulcer, abdominal pain), of cardiovascular disturbance (e.g. hypertension, ECG abnormalities) and fluid retention. Temperature, blood pressure, pulse and respiratory movements are monitored during drug administration, and signs of local irritation at the infusion site are looked for. Should the platelet count fall sharply, Billy will also be infused with pooled platelets.

The aims of chemotherapy are to remove the abnormal white cells, and to destroy the leukaemia stem cells (the lymphoblasts) in bone marrow. In so doing the 'crowding out' effect is removed and red cell and platelet counts in the blood are restored. Thus white blood cells will become 'normal', and the differential cell count will return to those observed in health (see Figure 21.1).

Parental education is an important part of this therapy. Billy will require considerable support because of the unpleasant effects of the drugs, the hair loss induced by chemotherapy, and the treatment regime. His parents will require support from the health-care team. Hence the emphasis on the management of side-effects and emotional support.

Acute lymphoblastic leukaemia responds well to treatment and has a high remission rate.

Further information

The success of chemotherapy partly depends on the aetiology of the leukaemia. The profound genetic disturbances noted in cancer cells can initially occur in a small number of cells as a result of a localized environmental 'insult' (cells produced from this focus will then dominate, and chemotherapy is directed at these cells). The alternative is that a genetic mutation could have been inherited or occurred *in utero*, in which case most of the stem cells will be affected. In this instance, ionizing radiation may also be used to eradicate the stem cells, and new, healthy cells introduced via a bone-marrow transplant.

Main reference

Clancy, J, and McVicar, A. J. 1995 *Physiology and anatomy: a homeostatic approach*. London, Edward Arnold.

22 The case of a child with growth-hormone deficiency

Theresa Atherton

Learning objectives

1 To explore the role of the endocrine system in controlling cellular/tissue homeostasis

2 To examine the impact of growth-hormone deficiency on tissue growth

3 To understand the rationale for care in relation to altered physiological processes

4 To understand the principles behind chemotherapy

Case presentation

Peter was born after a normal pregnancy. At three years of age, his mother began to express concern to her health visitor that something seemed to be wrong, as he didn't appear to be 'growing properly'. Soon afterwards the health visitor confirmed a downward trend in Peter's height and weight (the former affected to a greater extent than the latter). This seemed to support his mother's fears that Peter was getting fatter while not appearing to get any taller. Both parameters eventually fell below the third centile.

The GP who had been monitoring the situation, but had found no obvious cause for Peter's failure to thrive, decided to refer him to the local paediatrician for a second opinion. The paediatrician ordered extensive investigations which confirmed the diagnosis of growth hormone deficiency. Peter was subsequently referred to an endocrinologist for hormonal assessment.

Peter is now eight years old and has been having growth-hormone replacement under the supervision of the endocrinologist for the last four years. At the age of two, Peter was diagnosed as having asthma and has been treated with intermittent corticosteroid therapy.

Case background

Growth is a continuous, orderly and progressive process that follows the genetically predetermined trends of direction, sequence and pace.[1] Normal growth depends on many factors such as expected family growth patterns (which

All referals are from the main reference at the end of this case study.
[1] Refer to pp.577–9 for a description of growth patterns.

are genetically determined and culturally and environmentally influenced) and the general health and wellbeing of the child. However, individual children differ in their rate of growth and in their eventual size and capabilities, therefore it becomes increasingly more difficult to establish a satisfactory definition of what constitutes 'normal' or 'average' growth.

Anabolic hormones, in particular growth hormone (somatotrophin) secreted by the anterior and middle lobes of the pituitary gland,[2] promote physical growth. Growth hormone has widespread metabolic effects at cellular level. A reduction in the secretion of growth hormone will result in poor bone growth (in particular long bones) and an overall reduction in cellular metabolism. This might account for Peter's short stature and weight loss.

Growth-hormone deficiency does not bring about the same degree of weight loss as it does height loss. This results in an unequal weight/height distribution which might account for Peter looking 'fatter'. The weight gain might also be a side-effect of long-term administration of corticosteroids for the management of his asthma. The long-term administration of exogenous corticosteroids (synthetic steroids similar to glucocorticoid hormones produced by the adrenal cortex) help in the management of inflammatory disease processes, such as those found in asthma (see Chapter 8's case study on asthma). They operate by suppressing inflammation and promoting the repair of damaged cells and tissues; but, unfortunately, they are also known to suppress growth. The dose of steroids needed to induce growth retardation in children is not altogether clear and may vary from individual to individual, although doses in excess of 35 mg cortisone/metre squared of body-surface area are known to suppress growth (Buckler 1994). In Peter's case, the dose of steroids administered for the long-term control of his asthma were not thought to be of a sufficiently high dose to be responsible for his growth retardation. This fact

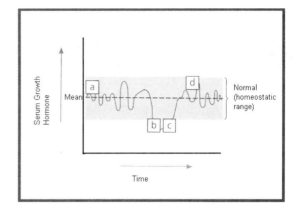

Figure 22.1 Growth–hormone deficiency. (a) Release of growth hormone appropriate for the age of the individual. Rate of growth within 'normal' limits . (b) Growth-hormone deficiency, resulting in a slowed rate of growth. (c) Commencement of growth-hormone-replacement therapy. Rate of growth increases accordingly. (d) Serum growth-hormone concentration restored to normal, appropriate for age. Restoration of growth rate to normal in children with uncomplicated hormone deficiency (as in Peter).

was also supported by the absence of features associated with high-dose steroid treatment, i.e. Cushing's syndrome.[3]

Care

Growth-hormone treatment is normally only prescribed where there is little doubt that there is a deficiency. Children with unambiguous growth-hormone deficiency, such as Peter, respond well to therapy (see Figure 22.1); whereas for children with more complicated pathologies, the outcome is less predictable. The management of Peter's growth retardation would be daily subcutaneous injections of growth hormone until such time as either an acceptable height is reached or his growth is almost complete. It would continue only as long as he maintained a reasonable height/velocity ratio (Buckler 1994).

---

[2] Refer to pp.439–41 for a description of the pituitary gland.

[3] Refer to p.450 and to Table 15.3 for the symptoms of cortisol hypersecretion.

There are four stages of growth assessment that will be essential in monitoring Peter's progress while he is on treatment: the measurements themselves, the recording of results, the interpretation of results and the subsequent action required.

An essential prerequisite for accurately monitoring growth is to have a standard against which a child can be compared. Such standards are depicted on growth-centile charts. This forms a convenient way of describing characteristic patterns of growth associated with the majority of children at periods where distinctive changes appear. (It was through the use of centile charts that Peter's growth retardation was initially noted and, by continuing this assessment process at frequent intervals, the improvement/deterioration in his condition can be monitored.)

It is imperative that measurements are accurately taken and recorded. Many factors can affect the latter. Faulty technique because of inexperience, faulty and/or ill-maintained equipment, inaccurate calibration, the incorrect positioning of child or equipment, the child being uncooperative, different people conducting the assessment, and measurements conducted at different times of the day may all predispose to inaccuracy.

To support adequate growth during childhood, nutritional intake[4] must exceed energy expenditure at an appropriate ratio. This ratio will vary in relation to growth patterns. Peter would have had a full dietary assessment during the investigative stage before diagnosis and then periodically to ensure that he continued to receive an appropriate nutritional intake balanced with his anticipated growth trajectory.

Further information

Physical growth does not rely solely on growth hormone. Thyroxine produced by the thyroid gland[5] is particularly active during childhood and is responsible for stimulating cellular metabolism, bringing about the ossification of bone and assisting in the formation of teeth.

Cortisol, a hormone produced by the adrenal cortex,[6] stimulates the conversion of fats and protein to glucose (gluconeogenesis) and is therefore involved in regulating blood-glucose levels alongside insulin, glucagon, somatostatin, thyroxine, corticosteroids and growth hormones.[7] Steroids released from the gonads also play a regulatory role in growth in later childhood.

There are many more wide and varied reasons why children fail to grow other than a growth-hormone deficiency. Such problems include an inherited/familial tendency to be short; malnutrition caused by an inadequate intake of food or an inability to take in food such as in chronic respiratory/heart disease; malabsorption and chronic inflammatory bowel disorders; psychosocial or emotional deprivation and anorexia nervosa; specific syndromes associated with short stature such as Down's and Turner's syndromes. Adequate assessment and investigation is therefore essential before any proposed course of treatment.

The medical management of growth-hormone deficiency is relatively simple but the impact of social, emotional and psychological difficulties arising in response to the abnormal growth status of a child must never be underestimated. Peter and his family will continue to require support during a course of treatment that will last for many years.

Growth hormone derived from human pituitary glands was banned in 1985 following identification of its causal link with Creutzfeldt-Jakob disease (Buckler 1994). The preparation is now synthetically produced.

[4] Refer to pp.111–17 for a discussion of the nutrient requirements during the lifespan.

[5] Refer to p.441 for a description of the thyroid gland.
[6] Refer to pp.443–5 for a description of the secretions of the adrenal glands.
[7] Refer to Figure 1.4 a homeostatic scheme for controlling blood glucose.

Main reference

Clancy, J, and McVicar, A. J. 1995 *Physiology and anatomy: a homeostatic approach.* London, Edward Arnold.

Other reference

Buckler, J. 1994 *Growth disorders in children.* London, British Medical Association.

23 The case of a baby girl with Hirschprung's disease

Stevie Boyd

Learning objectives

1 To revise the pathophysiology of the gut

2 To understand the physiological changes associated with Hirschprung's disease

3 To be aware of the treatment to relieve the symptoms

Case presentation

Ella was born at 38 weeks gestation, weighing 2.35 kg following an uncomplicated pregnancy. Her condition at birth was fair only with an Apgar score of 3 to 6 at 5 minutes, so she was transferred to the neonatal unit for observation. (The Apgar score is a numerical assessment of various aspects of neonatal wellbeing; a score of less than 7 indicates that support may be required. See Chapter 25's case study of the new-born 'well' baby).

Intravenous parenteral nutrition was commenced with 10 per cent dextrose, and prophylactic antibiotics were given.

Supplemental headbox oxygen was initially required to maintain oxygen-haemoglobin saturations above 92 per cent.[1]

At 24 hours of age, Ella's condition was much improved and she began enteral feeds of breast milk.

At 32 hours of age, Ella was reported to have had occasional non-bilious vomits and her abdomen was noted to be distended, with some periumbilical flaring. Two meconium plugs had been passed, followed by very small amounts of liquid stool. An abdominal X-ray revealed distended loops of intestine filled with gas. Oral feeds were discontinued and intravenous feeding was recommenced to maintain fluid and electrolyte balance. The monitoring of urine production and electrolyte content was also necessary to maintain homeostasis.[2] A nasogastric tube on free drainage was inserted to decompress the bowel and prevent further vomiting.

On day four, Ella's abdomen was still distended and becoming tender, so a surgical

---

All referals are from the main reference at the end of this case study.
[1] Refer to pp.307–9 for details of oxygen carriage by the blood.

[2] Refer to pp.96-7 for an overview of fluid and electrolyte balance.

opinion was requested. A provisional diagnosis of Hirschprung's disease was confirmed and a rectal biopsy revealed no ganglion cells. Arrangements were made for Ella to have a laparotomy, frozen section biopsies and the formation of a defunctioning colostomy to relieve her symptoms.

Case background

Hirschprung's disease was first described by Harald Hirschsprung, a Danish physician, in 1886, but it was not until 1948 that the pathological manifestations were understood. It is now a relatively common cause of intestinal obstruction in the new-born – 1:5000 live births, and is more common in boys (4:1 male:female). It is characterized by the absence of ganglion cells in Auerbach's (mysenteric) and Meissner's (sub-mucosal) plexuses[3] in the distal bowel from the internal sphincter and extending proximally for varying distances. The aganglionic segment always includes the rectum. The disease is confined to the recto-sigmoid colon in 75 per cent of cases; the sigmoid, splenic or transverse colon in 17 per cent; and the total colon is affected in 8 per cent.

Enteric ganglion cells are derived from embryonic vagal neural crest cells (the vagus is a major parasympathetic nerve). In the fetus, these first appear in the developing oesophagus at five weeks and migrate down to the ano-rectal junction by 12 weeks. These cells form the myenteric plexuses which are surrounded by longitudinal muscle. The submucous plexus is formed by the neuroblasts (poorly differentiated nerve cells) which migrate from the myenteric plexus across the circular muscle. The absence of ganglion cells in Hirschprung's disease has been attributed to failure of migration of the neural crest cells: the earlier the arrest of migration, the longer the aganglionic segment.

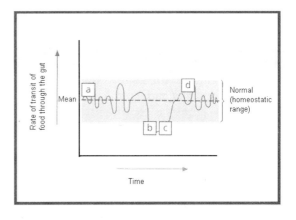

Figure 23.1 Hirschsprung's disease. (a) Rate of transit of food through the gut, mainly reflecting the levels of peristaltic contractions of gut segments. (b) Reduced transit because of obstruction. In Hirschsprung's disease, this is caused by segments of gut which lack the neural networks necessary for coordinating peristalsis (i.e. aganglionic segments). (c) Restoration of food transit through the gut. In Hirschsprung's disease, this is achieved in the short term by the removal of aganglionic segments and the formation of a colostomy at the distal end of the normally functioning segments of gut. (d) Re-established homeostatic functions: gut movement normal and food transit restored. In Hirschsprung's disease this is a long-term goal following the resection of the gut.

Dysfunctional adrenergic innervation has also been implicated.[4] Thus, the adrenergic nerves in the aganglionic bowel appear to be incapable of inducing muscle relaxation.

The aganglionic segment is non-peristaltic (see Figure 23.1), failing to relax, and this causes functional intestinal obstruction. Proximal to the aganglionosis, the normal bowel appears on X-ray as a megacolon with a dilated small bowel, above the distal obstruction. Abdominal distension, bile-stained vomiting, and failure to pass meconium are presenting factors in the new-born infant with this condition; foul-smelling diarrhoea also indicates the presence of enterocolitis. When a

[3] Refer to p.126 for a description of neural plexuses of the bowel.

[4] Refer to pp.386–91 for details of the autonomic nervous system.

gentle rectal examination is performed, the rectal wall always feels tight and resists further probing. This may cause the passage of meconium and flatus, followed by normal bowel movements for a few days or even weeks, before signs and symptoms of obstruction recur. Diagnosis is confirmed with suction biopsies of the colorectal mucosa, which is a simple and painless procedure.

Hirschprung's disease usually presents by 48–72 hours of age.

Care

Ella's clinical condition was stabilized while her parents were told of the necessity for surgery. Early diagnosis and the formation of a colostomy is the initial treatment of choice for neonatal Hirschprung's disease, and the operation should be carried out when a pathologist is available to report on frozen sections. The colostomy is usually made at the most distal part of normal innervated bowel, so several biopsies may be necessary. At operation, frozen section showed that the pathology stopped short of the descending sigmoid junction. The colon was divided, with the distal end left in the peritoneal cavity and the proximal end formed into an end colostomy.

Ella was ventilated for 18 hours to facilitate adequate gaseous exchange following general anaesthesia, and also to enable adequate pain relief with morphine without respiratory depression.[5] A nasogastric tube on free drainage remained *in situ* until bowel sounds were present.

Hydration was maintained initially with total parenteral nutrition (TPN) until four days post-operatively when full enteral feeds were tolerated. Ella was reluctant to bottle-feed on occasions and required tube-feeding. Weight gain was slow. Duocal was added to feeds to supplement the calorific intake.

Broad spectrum penicillin and gentamicin, plus metronidazole for anaerobic gut bacteria, were given prophylactically for three days.

The stoma nurse supervised care of the colostomy and early washout of the distal bowel to clear faecal retention. Skin integrity was preserved with meticulous management.

Family-centred care and early parental involvement with stoma care was introduced and encouraged.

Ella made an excellent recovery and was discharged at one month of age. She will hopefully have corrective surgery to her bowel at about six months. Between four and six months of age, further surgery to remove the aganglionic segment can take place. Currently, the preferred method is a Duhamel pull-through, via the retrorectal transanal approach, to bring the ganglionic bowel down to the ano-rectal junction where it is fixed to the distal rectum. The aganglionic bowel is resected at laparotomy. This is the operation that Ella would have. The original colostomy is sometimes closed at this stage; but depending on the ease of the anastomosis, a further, temporary, proximal defunctioning colostomy may be required for a month or so. Her family are aware that Ella's bowel management in the first three to four years after this operation will require understanding, patience and support, while normal function is restored.[6]

Further information

There is a family history of Hirschprung's disease in 3.6–7.8 per cent of cases. Unfortunately, genetic counselling is not universally available, as there is no clear pattern of inheritance.[7]

---

[5] Refer to Table 21.1 for the actions of analgesics.

[6] Refer to pp.150–1 for a discussion of functions of the large intestine.

[7] Refer to p.605 for an outline of polygenic inheritance.

The incidence of further anomalies associated with the disease can be as high as 35 per cent and range from neural crest anomalies, cleft palate, limb anomalies, regional anomalies of the urinary and intestinal tract and cardiac malformations. It is particularly associated with Down's syndrome.

Hirschprung's disease can be fatal in the absence of diagnosis. The main causes of death in undiagnosed Hirschprung's disease are sepsis from enterocolitis, and bowel perforation; early diagnosis and treatment are vital to decrease early mortality. Between 80 and 90 per cent of cases are diagnosed during the neonatal period.

Main references used

Clancy, J. and McVicar, A. J. 1995 *Physiology and anatomy: a homeostatic approach*. London, Edward Arnold.

Further reading

Clancy, J. and McVicar, A. J. 1997 Body fluid composition and cellular homeostates. *British Journal of Theatre Nursing* 7 (6).

Clancy, J. and McVicar, A. J. 1997 The principles of perioperative fluid replacement therapy. *British Journal of Theatre Nursing* 7 (8).

24 The case of a febrile toddler

Elaine Domek

Learning objectives

1 To revise the normal physiology of temperature regulation

2 To recognize the possible effect on a toddler of a high temperature

3 To explore the rationale for the care of a febrile child

Case background

Joe has had a febrile convulsion and has been brought to the hospital by ambulance. He is 18 months old, and has to date been a healthy baby, reaching all his normal milestones of development. Over the last 24 hours, he has become irritable, pulling at his ear and is not interested in his food. Rachel, his mother, is concerned by this. She took Joe to see their GP who diagnosed a middle-ear infection and prescribed antibiotics. While Joe's mother was at the chemist waiting for the prescription, Joe had a fit.

On admission to the accident and emergency department (A&E), Joe is no longer having a fit and is rapidly assessed using A/airway, B/breathing and C/circulation. Some secretions are suctioned from his mouth and nose and a set of baseline observations are taken. (These include weight, temperature, pulse, and respiration.) The nurse would have liked to have checked his blood pressure (B/P) but Joe became so distressed, screaming when she applied the cuff, that the attempt was abandoned. A child who is screaming will have a raised B/P because of the fight/flight response[1] as the child tries to extricate himself from the situation. The B/P in these circumstances will give no useful clinical information.

Joe's temperature is 40°C taken with a tympanic membrane sensor. This is a very accurate method of recording a child's temperature in seconds, rather than with other methods which take between four and six minutes and require a great deal of cooperation from the child. The temperature is recorded from the tympanic membrane (eardrum) and this method of temperature-monitoring is understood to be so effective because it records a temperature at a point in close proximity to the hypothalamus. The method is to gently tug the pinna of the ear to enable the canal to be straightened. This ensures a reading is taken from the membrane, not from the cooler walls of the ear canal. Joe's pulse and respiration are both elevated.

Joe is stripped to his nappy and a thin T-shirt with his mother's help. The doctor prescribes

All referals are from the main reference at the end of this case study.
[1] Refer to p.647 for a discussion of the fight and flight response.

rectal paracetamol and an antibiotic. These are administered in A& E, and Rachel is encouraged to get Joe to drink some squash before transfer to the ward for overnight observation. The nurse explains to Rachel about a febrile convulsion and gives her a handout which reinforces the main points.

Case background

Febrile convulsions affect 3–5 per cent of all children. They tend to occur between the ages of six months and three years, and will usually accompany intercurrent infections, typically viral illness, tonsillitis, pharyngitis and otitis media. They are unusual after five years of age. Otitis media is very common in early childhood and may often be the precipitating factor in febrile convulsions. This is caused by the shortness of the auditory tube which links the back of the nasopharynx with the middle ear.[2] The distance becomes greater as the child grows, and middle-ear infections become less frequent. In the infant and toddler, however, it provides a route for infecting agents to access the middle ear. The increased temperature of 40.0°C, recorded for Joe, is caused by the change in the set point in the hypothalamus.[3]

The normal set point in a toddler is 36.7–37.7°C but with infection the infecting pyrogens cause an increase in the normal set point (see Figure 24.1) in response to local reaction at the site of infection. The immaturity of the hypothalamus leads to the new set-point being too high (see the Introductory section for a discussion of set-points) and exciting a group of cells known as the epileptogenic focus. Neural tissue is most susceptible to a change in core temperature. This in turn leads to fits occurring at the onset of the fever (not after prolonged fever).

[2] Refer to pp.416–17 for a description of the anatomy and physiology of the middle ear.
[3] Refer to p.518 for a discussion of pyrexic responses to infection.

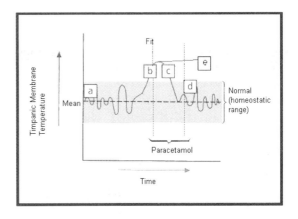

Figure 24.1 Febrile toddler. (a) Body temperature prior to febrile episode within homeostatic range of a toddler (36.7–37.7°C): balance between heat loss and heat gain. (b) A gradually rising temperature as heat gain exceeds loss, and a new set-point. If the new set-point is too high for neural function, a fit occurs. (c) Anti-pyretic administered as soon as possible together with other interventions (e.g. the removal of the toddler's clothes) to reduce body temperature. (d) Following antibiotic therapy the body temperature returns to its normal range. Over the next 24 hours the set-point is reset to the normal range. (e) Worst case scenario: no intervention takes place. Toddler is unable to rectify heat gain with heat loss. Temperature remains at the new set-point thus destroying neural tissue.

Care

1 Rapid assessment of airway, breathing and circulation (ABC) to check for respiratory or circulatory distress

2 Suctioning: infants and toddlers tend to nose breathe, so any collection of secretions will impair respiratory effort

3 Baseline observations of temperature, pulse and respirations are taken to facilitate the continuing monitoring for changes which may indicate that the child is in danger of having a further fit

TEMPERATURE

A continuing raised temperature following the administration of an antipyretic means that further fits may occur if other steps are not taken to cool the child. Removing clothes will disturb trapped air[4] and lessen the insulating effect that the clothes produce. Rectal paracetamol may be administered. It can be less distressing to a child to administer paracetamol this way, and it is more likely that the child will receive the full dose, especially when they may be resisting attempts to make them swallow a strange liquid in an unknown environment. The thin rectal mucosa means this can be a very efficient way to administer drugs: they are absorbed very quickly by this route if the rectum is empty.

RAISED PULSE AND RESPIRATORY RATE

Raised pulse and respiratory rate are evidence of the extra load that is being placed on these systems by a body that is having problems maintaining homeostasis related to the metabolic needs of cells. Cells which are generating heat will give off extra carbon dioxide. This will trigger a response in the peripheral chemoreceptors[5] to increase the respiratory rate. An increased pulse rate occurs with raised body temperature and in response to the increased respiratory rate.

The regular monitoring of vital signs will take place overnight, usually two-hourly, when oral paracetamol is used as it reaches its maximum effect in two-and-a-half hours. The children's nurse will also closely observe the child's behaviour and general condition of skin, to assess changes.

4   The child will be weighed and their weight will be recorded on their drug chart, so that drug doses can be accurately prescribed. Most preparations of drugs for children are in millilitres of solution: the number of millilitres necessary is multiplied by the weight of the child in kilograms to give a therapeutic dose.

5   Antibiotics have a role in stimulating antibody production.[6] In the case of a toddler, because of the risk of febrile convulsion, many doctors will prescribe them to help the child fight the infection.

6   The child will be encouraged to drink: squash is usually more palatable than water. The increased metabolic rate caused by the body's reaction to infection can easily lead to dehydration. This can be exacerbated by the large surface area in relative terms (roughly twice that of an adult) from which perspiration can take place. The child will suffer from physical fatigue following the fit, which can make them reluctant to drink.

7   Overnight observation is partly to observe the child and partly to support and reassure the parents who have undergone a very frightening experience. The time will also be spent in educating them in the steps they can take to lessen the chance of another febrile fit.

8   Reassurance is based on the fact that for many families the child will only have one episode of febrile fit. The likely occurrence of other febrile fits is linked to the age of the child (the younger the child the more likely they are to reoccur) as is a family history of febrile fits.

9   Education is centred around giving the parents the skills to recognize when the child is becoming pyrexial and the appropriate steps to take to minimize the risk of a febrile convulsion. These are:

• if the child appears hot when a hand is placed on the nape of the neck, or abdomen, it is advisable to remove outer layers of clothes to enable cooling to take place
• this should be closely followed by observation of the child for signs of pyrexia, such as red face, flushed skin, general lethargy

[4]Refer to pp.512–13 for a discussion of routes of heat loss.
[5]Refer to pp.312–14 for a discussion of chemoreceptors.

[6] Refer to pp.272–4 for a discussion of antibodies.

- administer an antipyretic such as oral soluble paracetamol in the recommended dose for the child's age, if their temperature appears to be increasing
- encourage a high fluid intake
- seek medical advice if fever persists

Further information

The seizures or fits are generalized being of a tonic/clonic nature (see Table 24.1) with the tonic phase lasting 10–20 seconds, and the clonic lasting about 30 seconds, but varying from a few seconds to 30 minutes.

Main reference

Clancy, J. and McVicar, A. J. 1995 *Physiology and anatomy: a homeostatic approach*. London, Edward Arnold.

Table 24.1 Examples of behaviour during fit

Tonic phase	Clonic phase
Tonic phase: lasts approximately 10–20 seconds	Clonic phase: lasts about 30 seconds
Manifestations Eyes roll upwards Immediate loss of consciousness If standing, falls down	**Manifestations** Violent jerking movements observed as the trunk and extremities contract and relax
Generalized muscle contraction with flexed arms and stiff body Legs, neck, and head extended An unusual piercing crying sound or may hold breath and become cynotic May have increased salivation	Soiling of urine and faeces may occur Gradually movements decrease in intensity The intervals become longer and cease altogether

25 The case of a new born 'well' baby

Elaine Domek

Learning objectives

1 To revise the physiology of temperature regulation

2 To recognize the consequences of a cold environment for a new born baby

3 To understand the rationale for the care of a new born baby to prevent hypothermia

Case background

Susan was born on a windy January day, by normal vaginal delivery, to Maggie, who has no known health problems, but an emergency in an adjacent suite caused the door to the suite to be in constant use. Once delivered, Susan was slow to breathe, so she was transferred to the resusitaire. At one minute Susan was pink, moving vigorously, grimaced on pharyngeal suctioning and was crying lustily with a healthy normal heart rate of 128 beats per minute (bpm). This gave an Apgar score of 9; at five minutes the score was 10.

The Apgar score is an internationally recognized method of assigning a mathematical number to a number of simple but rapid observations of the baby's condition in the first 20 minutes after birth (see 'Further information'). The scores indicate the baby's adaptation to extrauterine life and the need for intervention if the score is below 7.

Approximately 45 minutes later Maggie and Susan were transferred to the postnatal ward. Susan's temperature at this time was 36.8°C.

Her heartbeat was 132 bpm and respirations 30 per minute (pm). She weighed a healthy 3.2 kg. After an initial cuddle Susan fell asleep and was placed in a bassinet. These have high, clear plastic sides, so the baby can be observed but is protected from draughts. Soon Maggie was asleep. During this period the ward-cleaner moved Susan's bassinet near the draughty metal windows of the side ward.

On routinely checking the baby's temperature rectally three hours later it was found to be 35°C: this is hypothermic, as normal body temperature for Susan should be between 36.5 and 37.5°C. Both heartrate and respiration were slightly elevated at pulse 146 and respiration 36.

When Susan's low temperature was discovered, her blood was tested for evidence of hypoglycaemia. Her blood glucose at 2.0 mmol/l was within normal limits. Though very low for an older child, blood glucose of over 1.5 mmol/l is acceptable in term babies for the first 48 hours. After the blood test the distressed Susan was handed to Maggie to feed. Maggie was encouraged to place Susan close to her body so that radiant heat from her would be transferred to Susan.

After the feed Susan's temperature had risen slightly to 35.6°C. It was explained to Maggie that Susan's temperature was still low and that an overhead heater would be used over the bassinet and her temperature monitored. Susan's temperature continued to rise slowly with these measures and two hours later Susan's temperature was 37.2°C with a pulse rate of 134 bpm and respiratory rate of 32 pm. Mother and baby were discharged home the next morning.

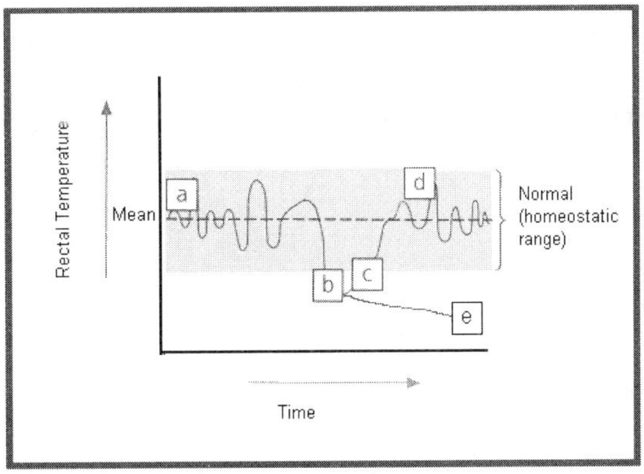

Case background

A new born baby can lose 0.25°C in one minute following birth, if the midwife fails to take precautionary steps to minimize heat loss. The main heat losses will be evaporative and radiant (Roberton 1993) because of the relatively large skin-surface area found in the neonate,[1] but both conductive and convective heat loss may also occur.

With a body temperature of 35°C Susan's metabolic processes will be acting to re-establish body-temperature homeostasis (see Figure 25.1). Neonates have no shiver reflex at this body temperature, therefore Susan will need to generate heat by non-shivering thermogenesis; this is thought to occur in the 'brown fat' which is located between the shoulder blades.[2] Brown fat is well served with blood vessels and it is the exothermic reaction of hydrolysis in the brown fat to free fatty acids and glycerol that will warm the blood as it flows through the brown fat.

Thermogenesis will increase the metabolic rate; consequently there will be an increased demand for oxygen. A phospholipid known as surfactant[3] is the lubricating solution which

Figure 25.1 New born 'well' baby. (a) Rectal temperature fluctuating within its homeostatic range. Heat loss and gain in balance. (b) Falling body temperature as heat loss exceeds gain. (c) Effective measures to correct the cause of heat loss in the infant. Non- shivering thermogenesis commences. (d) Return of rectal temperature to within homeostatic range. (e) Failure of physiological processes to restore body temperature, perhaps because of functional immaturity in the infant, or a persistence of a cool environment that maintains excessive heat loss.

provides surface tension that prevents the alveoli collapsing on expiration; this becomes thicker and less effective in cold temperatures. There is a risk that oxygen exchange which is already compromised by the immaturity of the respiratory system in the neonate will become even more difficult.[4]

The demand of the cells for more glucose to support non-shivering thermogenesis is increased at a time when the small stores of glycogen acquired during intrauterine life may have been depleted by the 'birth process'. The combined consequence of these two events is that it can quickly lead to a situation where the hypoxic baby becomes acidotic.[5] The normal pH for Susan would be in the range 7.26–7.49. If an acidosis is allowed to continue, the baby can suffer brain damage because of intracranial haemorrhage – and even death.

All referals are from the main reference at the end of this case study.
[1] Refer to pp.512–13 for a discussion of routes of heat loss.
[2] Refer to p.515 for a discussion of thermogenesis.
[3] Refer to p.300 for details about surfactant.

[4] Refer to p.303 for a discussion of gas exchange in the lung.
[5] Refer to p.58–60 for details of metabolic processes.

Care

This is mainly directed at reducing heat loss at the time of birth and until the baby is able to balance heat production with heat loss. This is known as thermoneutrality. This process develops during infancy.

At birth:

1 The baby should be delivered into a warm room, heated to 23–24°C, and every effort must be made to keep the room draught-free to prevent convective heat loss. Never uncover more than is necessary, to conserve heat, for example, when washing or bathing.

2 The baby should be wrapped immediately after delivery and dried as soon as possible to prevent the evaporative heat loss which will occur to a baby wet with liquor. Protect from draughts to reduce convective heat loss, and maintain a constant temperature of 21–24°C[6] in the nursery to reduce radiant and conductive heat loss.

3 The baby should be covered/dressed in dry clothes as soon as possible. This will insulate them and prevent radiant heat loss.

4 Any surfaces the baby is laid on should be covered, to prevent conductive heat loss.

5 Action is taken to clear and secure the airway. Suctioning is carried out if required to clear the mouth and then the nose, and oxygen is given to prevent hypoxia, if the baby is slow to breathe.

6 The baby is fed as soon as possible by breast or bottle, to prevent hypoglycaemia developing. Milk feed should be given frequently, two to four-hourly by breast or bottle depending on the baby's demands or requirements. A new born baby will need approximately 150 mls per kilogram in 24 hours. This total amount is divided, if bottle-feeding, by the number of feeds given during the period.

[6] Refer to p.515 for a discussion of the 'comfort zone' for ambient temperature.

Susan's care had been carried out correctly, but this highlights the importance of being aware of how draughts and low room temperature can affect young babies.

Further information

There are several ways to identify the infant who needs resuscitation. However, the Apgar scoring system remains the most useful in today's delivery rooms, even though it was first developed in 1953.

Table 25.1 The Apgar scoring system

Clinical features	0	1	2
Heart rate	0	<100	>100
Respiration	Absent	Gasping or irregular	Regular or crying lustily
Muscle tone	Limp	Diminished, or normal with no movements	Normal with active movements
Response to pharyngeal catheter	Nil	Grimace	Cough
Colour of trunk	White	Blue	Pink

Main reference

Clancy, J. and McVicar, A. J. 1995 *Physiology and Anatomy: A Homeostatic Approach*. London, Edward Arnold.

Other reference

Roberton, N.R.C. 1993 *A manual of neonatal intensive care*. London, Edward Arnold.

The case of a child with diabetes mellitus

26

Elaine Domek

Learning objectives

1 To revise the normal physiology of glucose metabolism

2 To consider the effects of hyperglycaemia on a child who is unable to secrete insulin

3 To explore the rationale for care of a child with diabetes

Case presentation

James is eight years old and has recently started drinking large amounts of sweetened drinks. James now complains of being too tired for football with his friends. His mother describes him as listless. His grandmother suggested the symptoms were similar to Aunt Sylvia's when she had insulin-dependent diabetes mellitus (IDDM) diagnosed as a child.

On visiting the GP's surgery, the doctor listens to the story. Knowing that diabetes often occurs in families, she checks James's blood glucose using a glucometer. This reveals he has a blood-glucose level of 12 mmol/L. She explains to James and his mother that Grandma may be right: the level is higher than normal (5–7 mmol/L). She is going to contact the paediatric diabetic nurse attached to the local paediatric department.

Case background.

Blood glucose normally remains within a homeostatic range of 3.5–5.5 mmol/L, but naturally rises after absorption of a carbohydrate-rich meal. In non-diabetics, insulin would be secreted from the beta cells of the islets of langerhans: a discrete cluster of endocrine tissue within the pancreas.[1] This stimulates the uptake of glucose in the hormone's target tissue: that is, the liver and skeletal muscle cells.[2]

A feedback mechanism is provided by the hypothalamus which stimulates the pancreas to secrete either insulin when glucose concentrations are beyond their homeostatic range, or glucagon (from alpha islets) when concentrations are below their homeostatic range. When blood is within its homeostatic range somatostatin (from delta islets) is secreted. This hormone has a paracrine role, inhibiting the secretion of insulin and glucagon.[3] In the IDDM patient there is a lack of, or below normal secretion of, insulin, so blood-glucose control is achieved by diet and insulin injections.

All referals are from the main reference at the end of this case study.
[1] Refer to pp.446–7 for a description of the pancreas.
[2] Refer to pp.58-9 for details of glucose metabolism.
[3] Refer to p.430 and p.446 for a definition of pancreatic secretions, and the role of somatostatin.

Glucose cannot be used by the cells in the absence of insulin, so protein is broken down in the muscles and converted to glucose (gluconeogenesis). This increases blood glucose but, still unable to use the glucose, the body recognizes its need for carbohydrates and creates a craving for sweet things. Often the metabolism of fats will also occur. This produces ketone bodies[4] used by cells as an alternative form of energy. Ketone bodies are strong acids that will lower plasma pH, resulting in ketoacidosis. Excess ketones are excreted in the urine and thus can be analysed. Ketones are also excreted by the lungs, causing in extreme cases an 'acid drop' breath. The 'acid drops' breath will be accompanied by deep sighing breaths as the body tries to correct the acid balance.

Care

Children are not admitted to hospital with suspected diabetes unless it is absolutely necessary. This is to maintain a 'normal' lifestyle for the child and their family. Children with diabetes are not 'ill' in the conventional sense, but suffer from a lack or deficiency of insulin so that normal cell metabolism can not take place.

Arrangements will be made for the diabetic nurse specialist to visit James and his parents at home that day and a leaflet prepared by the hospital would be given to James and his mother, which explains that diabetes is a disorder where the pancreas is not secreting enough insulin.[5]

In later visits the diabetic nurse educates James and his parents about diabetes in greater depth. She would show them the location of the pancreas using a flipbook of the body and explain that insulin is a hormone which is released into the blood when necessary to control blood glucose, so that the body always has the right amount of glucose available for its energy requirements.

She tells them that the causes of diabetes in children and young people are still not clearly

understood. We know that it can run in families, but there is no proven genetic link.

The nurse explains that glucose comes from the food we eat, so the levels are higher following meals but we always have some glucose in our blood and levels vary naturally throughout the day. Insulin is given by injection because it is inactive given by mouth. Using modern long-acting types of insulin, most people only need to inject twice a day. To check how well insulin is controlling the blood sugar, the test the doctor did can be done at home with the equipment supplied by the nurse. James's parents are not sure whether they like the idea of pricking James's finger but the necessity to do this is clearly explained. Good control means giving the correct dose of insulin to keep the blood glucose within the normal homeostatic range (see Figure 26.1). This is important because both short- and long-term complications can arise.

After some discussion on diet and that evening's meal, the nurse checked James's blood sugar and gave him a dose of insulin, as agreed with the consultant. Leaving James and his parents her phone number and some leaflets on diabetes to read, she departed promising to return in the morning to give James his insulin before breakfast.

A diabetic diet is a normal healthy diet. Special diets are not necessary, but a few simple rules have to be followed. These include:

- James will need to eat at regular intervals.
- Meals will need to consist of some long-acting carbohydrate, such as bread and cereals. These are broken down slowly, so that glucose is released slowly. This maintains blood-glucose levels within the homeostatic range longer. Short-acting carbohydrates like sweets cause a short burst of glucose.
- James will need insulin about half-an-hour before his breakfast and evening meal, to give the insulin time to take effect.
- When James takes part in any strenuous exercise, he will need extra carbohydrate beforehand: when our bodies work hard they require more glucose for energy.

[4] Refer to pp.59–60 for details of respiratory processes involving lipids and proteins.
[5] Refer to p.431, and pp.446–7 for details of insulin.

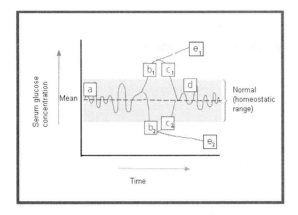

Figure 26.1 Diabetes. (a) 'Normal' serum-glucose concentration within a diabetic child as a result of good extrinsic control. (b_1) Hyperglycaemia caused by excessive carbohydrate intake. (b_2) Hypoglycaemia caused by poor carbohydrate intake, or increased utilization, as in exercise. (c_1) Effect of administered insulin to restore 'normal' serum glucose for the diabetic child . (c_2) Ingestion of biscuits or dextrosol to increase serum-glucose concentration for the diabetic child during hypoglycaemic episodes. (d) 'Normal' serum glucose restored by extrinsic control. (e_1) Persistent hyperglycaemia resulting from poor extrinsic control leading to long-term complications of diabetes mellitus. (e_2) Persistent reduced serum-glucose concentration leading to behavioural disorder or coma.

The nurse explains that if James needs more glucose his body will let him know. The signs may be feeling dizzy, starting to sweat, complaining of being hungry, having blurred vision or a trembling/tingling of his hands. These signs of low blood glucose are often referred to as a 'hypo' which is short for hypoglycaemia (low glucose). To counteract these symptoms some glucose is needed. This is available as dextrosol tablets, but sugar lumps or a sweet drink will also work. Once James feels better, the glucose should be followed by a long-acting carbohydrate.

For children with diabetes there are three main aims of treatment:

1 For James and his parents to be in control of the diabetes.

    The next day, the nurse continued to inform James and his parents about diabetes and how to use the blood tests to be in control of the condition. She also arranged an appointment at the children's diabetic clinic where they will meet the consultant, who specializes in children with diabetes, and a dietician, who can give further advice on meals, tailored to James's preferences. By knowing about diabetes and who is available to support them, James and his parents can regain control of their lives.

    During the visit James's mother asked why James had been so thirsty and listless. The nurse explained that diabetics who have recently acquired the condition are thirsty because the amount of glucose in the blood is above the normal range (that is, the renal threshold is exceeded). So glucose appears in the urine, causing more urine to be passed (polyuria). Since this urine is very dilute it triggers a homeostatic response, in relation to the loss of fluid,[6] and the anti-diuretic hormone is secreted from the posterior pituitary causing a feeling of thirst (polydipsia).

2 For the child to lead a normal life participating fully in all aspects of childhood, play, education and social activities.

    The nurse offers to contact James's school so that he can return knowing that his teachers understand the modern management of diabetes, and so they can express any individual concerns. She also offers information about the British Diabetic Association local branch and will arrange for James and his parents to meet other families for mutual support if they so wish.

3 To minimize the short- and long-term complications associated with high or low blood-sugar levels. Short-term complications can be:

    • attacks of hypoglycaemia (low blood sugar) which are usually related to too little food, too much insulin, and more exercise than usual. This promotes behavioural disturbances and, in the extreme, possibly coma

[6]Refer to pp.338–9 for a discussion of water homeostasis.

- attacks of hyperglycaemia (high blood sugar) which are usually related to too many foods containing short-acting carbohydrates, too little insulin, and not enough exercise.

In the long-term major complications can occur such as pathological changes to the microcirculation (caused by proteins being trapped, because of the sticky glucose molecules: glycosyl, atheroma deposits, etc.) which can lead to blindness, renal failure, skin ulcers, myocardial infarction, and hypertension. Other problems include urinary tract infection and polyonephritis, excessive weight loss, and the loss of sensory innervation especially from the extremities .

The management of diabetes in children is therefore a balancing act between meeting their energy requirements with a healthy diet and providing enough insulin so that the cells can use the glucose provided.

Further information

Two per cent of the population have diabetes (approximately 750 000 people). Of these approximately 180 000 will have insulin dependent diabetes mellitus (IDDM).

Main reference

Clancy, J. and McVicar, A. J. 1995 *Physiology and anatomy: a homeostatic approach*. London, Edward Arnold.

The case of a baby with bronchiolitis

27

Elaine Domek

> *Learning objectives*
>
> **1** To revise the normal physiology of the respiratory system
>
> **2** To consider the altered physiology in bronchiolitis
>
> **3** To explore the rationale for care of the baby with bronchiolitis

Case presentation

Sarah and her mother, Becky, are admitted to the children's ward in early December. Sarah has a three-day history of a cold, which has progressively led to poor feeding.

Sarah is Becky and John's first child, and is ten weeks old. She was born in late September at 36 weeks gestation by normal vaginal delivery. Sarah weighed 2.5 kilograms and despite being early she was a healthy new born with none of the problems associated with prematurity. Becky and Sarah were discharged after 48 hours. Becky decided to bottle-feed Sarah, and everything had been going well. There were no problems at the six-week-check. Sarah was feeding well and gaining weight. One week ago at the clinic Sarah weighed 3.7 kilograms.

On admission Sarah had a breathing rate of 45 breaths per minute (tachypnoea), a pulse of 160 beats per minute (normal 100–220 bpm), and a temperature of 37.2°C. Becky had given her oral paracetamol suspension three hours before admission and Sarah's oxygen saturation was 92 per cent. Oxygen saturations are recorded by a pulse oximeter which records the arterial oxyhaemoglobin levels at a peripheral site such as finger, toe or earlobe. Her weight was 3.2 kilograms. Sarah had a runny nose and an irritating cough. The provisional diagnosis was bronchiolitis (caused by the respiratory syncytial virus (RSV). This disorder tends to occur in annual epidemics, especially in the winter months. Diagnosis is often on clinical signs, but it may be confirmed by X-ray and the cultivation of secretions from nasopharyngeal aspirations, often referred to as NPA. specimen. Recovery usually occurs within seven to ten days.

Case background

There two main problems with bronchiolitis:

1 Dyspnoea

The respiratory syncytial virus affects both children and adults. The first symptoms are those of a cold; most adults are unaware they have RSV. Young children, especially under three years old, seem to develop bronchiolitis, in which the bronchioles[1] become inflamed. This reduces the size of the airways, causing an obstruction. Those particularly affected are premature babies, and babies with circulatory or other respiratory problems. As a general rule the babies admitted to hospital are usually under one year of age.

2   Dehydration
This occurs because:

- young babies obtain both nutrients and fluids from milk feeds[2]
- the increased respiratory effort used by the infant to breathe leaves them too exhausted to suck. This can be overcome by using a nasogastric tube to administer the feed
- a full stomach interferes with the use of abdominal muscles to breathe. This can lead to vomiting after feeds which causes a deficit in fluid balance
- fever and an increased metabolic rate increases fluid requirements
- an increased respiratory rate will increase the one-third of insensible fluid loss[3] that occurs through the respiratory tract

(Please see Chapter 20's case study of a child with pyloric stenosis for a discussion on dehydration in infants.)

Care

Matthew, their named nurse, explains that Sarah will be barrier-nursed to prevent the spread of infection as the virus is transmitted through contact with infectious respiratory secretions. Sarah and Becky are placed in a family room, with a bed for Becky and a cot for Sarah. The importance of hand-washing,

especially before leaving the room, is explained to Becky, and why it would be better if Becky did not visit any other rooms.

Matthew explained that Becky or John were very welcome to stay with Sarah and that close family or friends could visit at any time. It was also explained that where possible the nursing staff would keep to Becky's routine and that she would be encouraged to care for Sarah, with help if necessary, but that if she felt tired or needed a break the nurses were there to take over.

Matthew then tells Becky that the doctor will be coming to see Sarah soon, and that there were two main problems to be addressed. The first was to help Sarah breathe more easily. Matthew then gently suctioned Sarah's nasal passageways and throat, explaining to Becky what he was doing and why. Sarah was placed in a baby chair, which Becky said she usually enjoyed.

Becky was very concerned about Sarah's feeds as she had not completed a whole feed for the past 24 hours and at the last feed she had only managed 40 ml instead of 90 ml. Matthew explained that this was fairly normal and this was the second problem. Because they are small, babies can easily become dehydrated. The doctor would probably insert a cannula into the vein for the administration of intravenous fluids if needed. A nasogastric tube would also be used to ensure a full feed. Although slightly reassured that all this would help Sarah, Becky was concerned that it might cause her pain.

Matthew agreed that neither procedure would be very pleasant, but they would try to minimize any discomfort and Becky was very welcome to stay with Sarah when they were done. He explained that by being there and talking to and comforting Sarah, she could help to minimize the discomfort. If she did not wish to be there, one of the nurses would take care of Sarah.

Matthew then explained that all Sarah's paperwork would be kept outside the room, and certain things would be recorded frequently. These would include the observations of Sarah's temperature, pulse and respiration (TPR) and oxygen saturations (see

---

All referals are from the main reference at the end of this case study.
[1] Refer to pp.294–7 for an overview of the anatomy of the respiratory system.
[2] Refer to p.115 for details of milk composition.
[3] Refer to pp.96–7 for a discussion of sensible and insensible water loss.

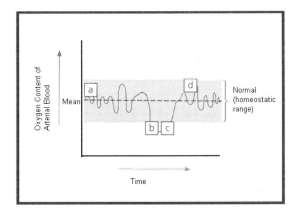

Figure 27.1 Bronchiolitis. (a) Oxygen content of arterial serum within the homeostatic range necessary to support tissue functions. (b) Decreased arterial oxygen content (hypoxaemia). In bronchiolitis this is a result of alveolar hypoventilation arising because of inflammation of the bronchioles. (c) Implementation of therapies. Alveolar ventilation is promoted by ideal posture and by the removal of airway secretions. Alveolar composition is further enhanced by the use of oxygen therapy. (d) Restoration of serum oxygen content to normal .

Figure 27.1) as well as the amount of feed she takes and the number of wet and dirty nappies she has. He asked Becky if she would let the nurses know when she changed and fed Sarah.

When Sarah was seen by the doctor, it was agreed to insert a nasogastric tube. The intravenous (IV) cannula was inserted and her mild dehydration was treated overnight with IV fluids. Sarah was started on oxygen at 40 per cent concentration, while sitting in the chair with a perspex cover added, if her saturations dropped below 92 per cent.[4]

Sarah settled well in the chair, and initially her oxygen saturations improved. But overnight they dropped to the high eighties and an apnoea monitor was put on. Oxygen was administered as prescribed. Unable to complete her feeds, the nasogastric tube was used to ensure Sarah received her full feed over the next two days. Frequent suctioning of nose and mouth was carried out by the nursing

staff and three days later Sarah was ready to go home.

The measures taken to relieve dyspnoea are:

1   Place the baby/child in a position which will aid respiratory effort. This can be in a chair or a tilted cot. This enables chest expansion to be as full as possible, allowing the full use of accessory muscles of respiration.[5]

2   Ensure that the naso/oral passages are clear, by use of gravity and suction if necessary.

3   Maintain frequent close observation of the child's temperature, pulse and respiration (TPR), and O^2 saturations, decided by child's condition. For a discussion on the importance of TPR please see Chapter 24's case study on the 'febrile toddler'. Monitoring oxygen saturations continuously or intermittently can provide evidence of the efficacy of gaseous exchange.[6]

4   To compensate for falling oxygen saturations or in response to poor oxygen saturation, oxygen is administered via a headbox or nasal cannulae. This should be administered warm and moist, as cold gas can cause further constriction. Babies become hypoxic very quickly because of their normal, fast metabolic rate.

5   Because of the risk of respiratory arrest in a 'sick' baby, or one admitted with a history of apnoea, these children are often nursed with an apnoea monitor *in situ*. This will alert nursing staff to potential problems if the baby is left alone in a cubicle.

6   Apnoeic attacks or increased respiratory effort demonstrated by tachypnoea, recession, or falling oxygen saturation, may indicate the need for respiratory support.

[4] Refer to pp.307–9 for a discussion of oxygen carriage by the blood.

[5] Refer to pp.298–9 for a discussion of respiratory muscles.
[6] Refer to pp.303–4 for a discussion of gas exchange in the lung.

They can be evidence of the child's failure to compensate for the inadequate oxygen exchange taking place in the alveoli because of the inflammation of the bronchioles. This leads to increased carbon dioxide levels.[7] The baby's efforts to remove the carbon dioxide by increased respiration and extra respiratory effort involving the accessory muscles of respiration leads to exhaustion of the respiratory and cardiac muscles, precipitating respiratory/cardiac arrest.

Because a child's heart is usually healthy, children are likely to suffer respiratory arrest first. Once respiration has faltered, cardiac arrest will follow unless prompt respiratory support is offered. This can include oxygen administration by continuous positive airway pressure (CPAP) or support on a ventilator.

Further information

Carey and Clare (1991) stated that the respiratory syncytial virus is so common that 100 per cent of the population will have been exposed to it by the age of two years. In adults and older children, a cough and runny nose are often the only discernible symptoms. These appear no different than a 'slight' cold, but it is in the under twos with their immature respiratory systems that the problems associated with bronchiolitis are most often prevalent.

Main reference

Clancy, J. and McVicar, A. J. 1995 *Physiology and anatomy: a homeostatic approach.* London, Edward Arnold.

Other reference

Carey, M. A. and Clare, E. R. 1993 Nursing planning, intervention and evaluation for altered respiratory function. Cited in Jackson, D. and Saunders, R. *Child health nursing.* Philadelphia, Lippincott.

---

[7] Refer to pp.312–14 for a discussion of the role of carbon dioxide in regulating lung function.

The case of a boy with hypersensitivity reaction: asthma

28

Helen Thomas

> ## Learning objectives
>
> **1** To identify the role of the immune system in hypersensitivity reactions with specific reference to humoral and cell-mediated activities
>
> **2** To explain the possible causes of asthma and the subsequent patho-physiological changes that occur within the airways that characterize asthma
>
> **3** To demonstrate an understanding of pharmacological intervention in the treatment of asthma in young children
>
> **4** To justify the importance of education in the on-going management of individuals with asthma

Case presentation

Thomas is four years old. He has a long-standing history of recurrent chest infections and persistent cough. More recently, he has developed a noticeable wheeze which gets worse at night and continues during the morning at nursery school. These nocturnal episodes have been causing interruptions to his sleep and he has had to miss school on several occasions because of a persistent cough.

The family GP has taken a full clinical history from Thomas's mother. The history shows that in addition to the symptoms of persistent cough and nocturnal wheeze Thomas also has a history of mild eczema and allergies to certain foods such as eggs and peanuts. His father has a history of asthma from the age of 12 years.

Following a physical examination the GP referred Thomas for further investigations:

- Allergy skin (Heaf) tests demonstrated a positive skin test to airborne allergens.[1]
- Peak expiratory flow (litres/minute) recordings were significantly below the predicted values for age and height, indicating an increased airway resistance.[2]

All referals are from the main reference at the end of this case study.
[1] Refer to pp.287–8 for details of the immune response to allergens.
[2] Refer to p.302 and to Figure 11.7 for details of the measurement of airway resistance.

- Forced expired volume in one second (FEV$_1$) was reduced in comparison to predicted values for age and height, indicating increased airway resistance.[2]

Case background

Once asthma has been recognized and diagnosed the aims of care are to abolish its symptoms and restore normal or best possible long-term airway function. The risk of severe attacks must also be reduced and in children normal growth and minimizing absence from school are priorities.

Asthma is the most common chronic illness in childhood. It is a disease of the airways characterized by chronic inflammation with infiltration of lymphocytes, eosinophils and mast cells[3] together with epithelial desquamation, thickening and disorganization of the tissues of the airway wall, and mucus-plugging of the airways. Such pathology presents as variable airflow obstruction associated with the inflammation of the airways and the symptoms of cough, wheeze, chest-tightness and paroxysms of dyspnoea. The excessive airway-narrowing occurs in response to a variety of provoking stimuli such as allergens, environmental pollutants, exercise, infections, drugs and psychological factors amongst others.

Hypersensitivity reactions[4] are an exaggerated response to an antigen (allergen) following previous exposure. The resulting antigen/antibody reaction induces the release of large quantities of chemicals, enzymes and cell-stimulators. This response by the immune system is often associated with humoral immunity reactions.

In forced expirations, an individual should be able to expel 80 per cent plus of the vital capacity of the lung in one second. Peak flow rates and FEV$_1$ tests are simple means of monitoring the extent of airway-resistance changes.[5]

Care

Thomas was initiated on a trial of therapy for asthma (based on The British Guidelines on Asthma Management 1995). This included:

- an intermittent inhaled short-acting beta-2 adrenoreceptor agonist such as salbutamol via an appropriate dose-metered inhaler to induce bronchodilatation and large volume-spacer as required. These drugs affect the actions of the sympathetic nervous system on airways.[6] Large volume-spacers are devices which particularly help children to administer their inhaled drugs effectively, as the synchronization of intake of breath with the operation of an inhaler device is not required. They also help to reduce the unwanted side-effects of inhaled steroids, such as a fungal infection of the oral cavity known as 'thrush'. Large volume-spacers increase the lung deposition of the inhaled drug and are easier to use than some inhalers for smaller children. Instruction on the correct use of inhalers and peak flow meters is vital and evaluation of technique must be done regularly
- regularly inhaled sodium cromoglycate 10mg three times daily via an inhaler and large volume-spacer. The regular inhalation of sodium cromoglycate can reduce the incidence of asthma attacks and is useful in children as it may prevent the need for steroid therapy. The mode of action of sodium cromoglycate is not completely understood, but it is known to reduce the inflammatory response (i.e. histamine release) to irritants and as such it is an important preventive drug

---

[3] Refer to pp.178–81 for details of white blood cells.
[4] Refer to pp.287–9 for a discussion of hypersensitive responses.

[5] Refer to pp.298–9 for an explanation of the need of low airway resistance.
[6] Refer to pp.386–91 for a discussion of autonomic receptors and the actions of the sympathetic nervous system.

- home peak flow monitoring: instruction on technique/recording to ascertain the severity and progression of changes in airway resistance
- a symptom diary card to monitor the frequency of episodes and the allergen type
- education: information is given about the disease process, the causes, treatment and self-management programme. Ongoing support for the family via the nurse-led asthma clinic or GP

This trial was undertaken aggressively for a three to four week period since most childhood asthma would be expected to show some response to pharmacological therapy within that time.

If Thomas's asthma has not stabilized after this period of time an inhaled steroid or a short course of prednisolone tablets may be required as a trial for one month. Steroids have anti-inflammatory effects. A reassessment of peak flow recordings and symptom control from the diary is necessary to determine further treatment. Indicators of assessing the outcome of asthma treatment include the number of days off school, amount of day time and night time cough and wheeze, limitation of activity, frequency of relief medication, inhaler and peak flow technique and the understanding of parents or child that medications must be varied according to symptoms or peak flow recordings or both.

Successful management in children involves participation from both the parents and the child, in partnership with the health professional. Education is the foundation for long-term successful management as it fosters understanding and effective skills. Appropriate changes in behaviour are more likely to occur if the child and family are given an adequate opportunity to express any fears or concerns, and expectations of both the condition and its treatment. Much can be achieved by the identification and subsequent avoidance of known triggers.

The combination of pharmacological and educational interventions should lead to the effective long-term management of asthma (see Figure 28.1).

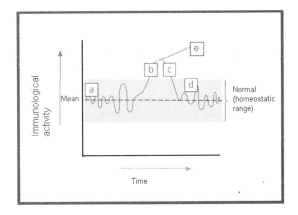

Figure 28.1 Asthma. (a) Normal homeostatic sensitivity and functioning of the immune system. (b) Hypersensitive reaction by the immune system. In asthma this is a response to frequent exposure to an irritant, and is referred to as a Type I response. (c) Intervention to reduce the reaction. In the short term, this will involve pharmacological therapy. In the long term, avoidance of the irritant responsible will help to reduce the reaction. (d) Restoration of normal immune system activity. (e) Uncontrolled hypersensitive reaction. This is potentially life-threatening as there will be pronounced lung hypoventilation and resultant hypoxia.

Further information

Inhaled allergens are the most common route for precipitating atopic (allergic) asthma especially in children. Inhaled allergens include pollens from grass, trees and weeds, fungi, the house-dust mite and animal dander. Pollens and fungal allergens tend to cause seasonal symptoms of allergic rhinitis and/or conjunctivitis. Allergy to house-dust mites (dermatophagoides pteronyssinus or dermatophagoides farinae) are extremely common. The former is the most common allergen causing IgE-mediated hypersensitivity[7] reactions in asthma, especially in children. Asthma triggered by animal dander is most commonly associated with cats.

A strong family history of allergy (atopy) or asthma is often associated with children who

[7] Refer to pp.272–3 for an overview of antibody types.

develop persistent asthma. Allergies to foods and/or the presence of atopic dermatitis at an early age indicate the presence of an atopic immune system. Later sensitivity to airborne allergens mediated by IgE antibody are implicated in the development of asthma. It is feasible that sensitization at a cellular level, for example, the subset of T-lymphocytes necessary for the production of IgE antibody by B-lymphocytes may even pre-empt the development of IgE antibody to airborne allergens constituting a cell-mediated hypersensitivity reaction. T-cells may be capable of producing the cytokines responsible for activating eosinophils in the airways.[8]

Main reference

Clancy, J. and McVicar, A. J. 1995 *Physiology and anatomy: a homeostatic approach*. London, Edward Arnold.

Other reference

The British Guidelines on Asthma Management 1995. Review and Position Statement: Thorax, February 1997, Volume 52, Supplement 1.

---

[8] Refer to pp.275–8 for a discussion of cell-mediated immunity.

The case of a child with nephrotic syndrome

29

Theresa Atherton

Learning objectives

1 To revise the physiology of urine production (filtration)

2 To revise the role of plasma proteins, in particular albumin, in the physiology of tissue-fluid production and reabsorption

3 To revise the related homeostatic mechanisms for the maintenance of blood pressure

4 To understand the rationale for care given in relation to the disordered physiology

5 To understand the principles behind chemotherapy

Case presentation

Rosie is three years old: the only child of Graham and Gill. During the last two weeks, Gill has noticed that Rosie appears to be rapidly gaining weight and has 'puffy' eyes when she gets up in the morning. On close questioning it appears that Rosie's urine output has diminished and its appearance is dark and frothy. Rosie seems to have lost her appetite and on examination, her abdomen is swollen. A usually active and inquisitive child, Rosie is now lethargic and complains to her mother that her tummy hurts. Gill also states that Rosie has had a tickly cough for the last three weeks. Rosie is subsequently diagnosed as exhibiting the symptoms of nephrotic syndrome.

Case background

Nephrotic syndrome is the most common presentation of glomerular abnormality in childhood (Campbell and Glasper 1995). The syndrome is attributed to an increased permeability of the glomerular membrane to macromolecules,[1] in particular the plasma protein albumin, which results in large quantities of urinary protein loss (proteinuria and hypoproteinaemia) (Figure 29.1). The majority of subsequent symptoms can be linked directly to this primary abnormality.

All referals are from the main reference at the end of this case study.
[1] Refer to pp.325–6 for a description of the glomerulus and its role as a filter.

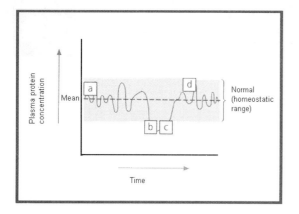

Figure 29.1 Nephrotic syndrome. (a) Serum protein concentration fluctuating but remaining within homeostatic norms. (b) Reduction in serum-protein concentration (hypoproteinaemia) as a result of synthesis being less than protein metabolism/excretion. In nephrotic syndrome, the problem is one of protein loss in the urine: the rate of loss is in excess of the rate of synthesis by the liver. (c) Return of serum-protein concentration toward normal, as protein excretion in nephrotic syndrome decreases and protein synthesis continues. (d) Complete restoration of serum-protein concentration to within homeostatic norms. Symptoms of nephrotic syndrome will usually reverse as this occurs.

Plasma proteins play a major role in maintaining colloidal osmotic pressure in the blood capillaries.[2]

In nephrotic syndrome, a significant reduction in plasma protein results in a homeostatic imbalance where fluid is retained in body tissues causing oedema.[3] This phenomenon could account for Rosie's rapid weight gain which is greater than that which would normally be expected for a child of this age. The problem is exacerbated by the abnormally high retention of sodium and water by the kidneys.[4]

Oedema is most prominent following periods of inactivity where fluid accumulates in tissues under the influence of gravity. This is referred to as 'dependent oedema'. In Rosie's case this might account for her puffy eyes on waking. Dependent oedema disappears as the person mobilizes and it would be expected that the puffiness around Rosie's eyes would reduce as the day progresses.

Tissue fluid can also accumulate in body cavities such as the peritoneal cavity, giving rise to a condition called ascites; or, more commonly, it can cause local swelling and engorgement of abdominal organs. This might account for Rosie's swollen abdomen: her anorexia and pain are also attributed to this problem.

Pronounced oedema also results in a lack of circulating blood volume (hypovolaemia), which reduces the amount of available blood entering the nephron at the beginning of the filtration process and results in oliguria. In severe cases anuria and renal failure may be a feature.

Plasma proteins (prothrombin, fibrinogen and gamma globulins) are also involved in body defence[5]. In nephrotic syndrome, an excessive loss of plasma protein may put the child at risk from an increased bleeding time and infection. The fact that Rosie has had a tickly cough for the preceeding three weeks may mean that her body defences are unable to overcome the infection. She may also be at risk from secondary infections while her illness persists.

It might be expected that hypotension would be a feature of the syndrome, but classically blood pressure remains normal or only minimally affected. Specialized cells in the walls of the afferent arterioles of the kidney (juxta-glomerular apparatus) recognize the reduction in renal blood flow and, as a homeostatic controlling device, they release the enzyme renin which converts the plasma precursor angiotensinogen to angiotensin.[6] Angiotensin acts as a powerful vasoconstrictor which brings about a compensatory change in arterial blood pressure. Angiotensin also stimulates the secretion of

---

[2] Refer to pp.92–4 for details of tissue-fluid formation.
[3] Refer to pp.89–91 for a description of body-fluid compartments.
[4] Refer to pp.339–40 for details of sodium and water homeostasis.

[5] Refer to pp.183–5 and to Table 10.1 for details of the involvement of plasma proteins and gamma globulins in blood-clotting and in defence reactions respectively.
[6] Refer to pp.239–40 for discussion of the role of the kidneys in blood-pressure homeostasis.

aldosterone which brings about further salt and water retention by the kidneys.[7] While helping to maintain circulating blood volume, the salt and water retention brings about a potential worsening of oedema.

Care

During the oedematous phase of her illness, Rosie is restricted to activities in and around her immediate bed area as she may become easily fatigued. Oedematous tissues are prone to breakdown; so while resting in bed, Rosie's skin is inspected every two hours for signs of pressure-sore development and she is encouraged to change position. Once the oedema has subsided Rosie will be allowed to resume her normal activities.

Should Rosie's plasma albumin concentration fall to a sufficiently low level that severe hypotension and circulatory collapse become an imminent danger, it may be necessary for her to have a transfusion of plasma or its substitute.

Rosie's urine is tested for protein (to assist the monitoring of plasma albumin loss) and she is weighed daily at the same time and in similar clothes (to monitor the degree of fluid retention/loss during the treatment phase).

A balanced diet will be encouraged, with extra protein supplements being added should Rosie suffer actual weight loss as opposed to fluid-weight loss. Further exacerbation of sodium retention will be prevented by Rosie avoiding foods which have a high salt content.[8] Fluids will be given as normal. A record of all fluid input and output is made.

Current infections will be treated with antibiotics. These will also serve as prophylaxis for secondary infections.

Corticosteroids are the primary therapeutic agents in the management of nephrotic syndrome. Medication will be started from the moment Rosie is diagnosed and will usually continue until her urine is protein-free for a period of approximately 10 to 14 days. If Rosie responds to the corticosteroid treatment she should have a diuresis in approximately 7 to 21 days, following which her fluid balance should return to normal, the diuresis coinciding with a reduction in the excretion rate of protein in the urine. Rosie's remaining symptoms should all be expected to diminish at this time.

Further information

Although the syndrome most commonly develops as a primary idiopathic disease (80 per cent of cases), it may also develop as part of an existing systemic disease process, or as a congenital abnormality. The prognosis for complete recovery is generally good for all children falling within the idiopathic category who respond favourably to steroid therapy but there is a risk of relapse which decreases with age (Campbell and Glasper 1995). Children who require frequent courses of corticosteroids because of relapse are at serious risk of side-effects which may affect growth and development as well as their general health. Such children may be treated with immunosuppressants such as cyclophosphamide or cyclosporin A. Unfortunately, serious side-effects also accompany the use of such drugs. Careful and thorough counselling of both child and family is advocated before any course of chemotherapy.

Main reference

Clancy, J. and McVicar, A. J. 1995 *Physiology and anatomy: a homeostatic approach*. London, Edward Arnold.

Other reference

Campbell, S. and Glasper, A. (eds) 1995 *Whaley and Wong's children's nursing*. London, Mosby.

[7] Refer to pp.444–5 for information about the control of aldosterone release.
[8] Refer to pp.97–8 for a discussion of the link between sodium chloride and extracellular fluid volume.

PART THREE

Mental health case studies

The case of an adolescent girl with anorexia nervosa

30

Gibson D'Cruz

Learning objectives

1 To understand the role of a sufficient calorie intake (in the form of a normal 'balanced' diet) on cellular and consequently tissue and organ function

2 To appreciate the factors that determine the minimum calorie requirements for an individual

3 To recognize that the onset of anorexia nervosa is associated with social, psychological and physiological factors

Case presentation

Jane Smith is a 17-year-old woman who was admitted, at the request of her GP, to the acute ward of a large psychiatric hospital. On admission to the ward, Jane was noted to have a pale complexion and looked emaciated and thin. There was also an apparent lack of subcutaneous fat, and parts of her skin – especially her face – was covered with very fine hair. She reported that she had difficulty in getting to sleep and, when sleep occurred, it was only for short periods of up to an hour. She also tended to wake up earlier.

During the nursing assessment, Jane denied that she was experiencing any medical problems. She appeared aloof and was restless – always in motion during the interview. On further questioning, Jane reported that she had not menstruated for the past six months and denied any possibility of being pregnant. Although Jane expressed feelings of sadness and depression, she did not express any desire to commit suicide. Jane felt that, although to others she may appear thin, she was fat and needed to lose more weight so as to appear attractive. Jane occasionally felt hungry but she managed to suppress the desire for food by diverting her thoughts to images of the fashion models who were featured on television and in the weekly magazines to which she often subscribed.

Jane's mother, who accompanied her to the ward and was present during the interview, said that Jane had been very careful with her dietary intake for a long time. Jane used to eat only very small amounts and occasionally missed entire meals. Jane was always aware of the amount she had eaten during the course of the day. Mrs Smith suggested that the reason for this was that Jane had constantly expressed a desire not to appear fat either to herself or to her peers. Mrs Smith also reported that if comments were made about Jane's dietary habits, Jane was likely to become hostile and if anyone encouraged Jane to eat more, she would burst into floods of tears and abruptly leave the room. Mrs Smith said that Jane had become withdrawn and seemed very preoccupied with herself and this had only begun since she started taking an active interest in her eating patterns.

Mrs Smith reported that although Jane was not interested in her own food consumption, she helped to prepare meals for the rest of her family. She took pride in ensuring that the meals she prepared were well presented and were enjoyed by all who ate them.

Mrs Smith commented that until Jane began to be very concerned about her food consumption, she had been a happy girl and was always striving for perfection in everything she did. She took part in games and enjoyed dancing and for a short period of time had ballet lessons. Even after she began dieting, Jane constantly did physical exercise and it was only in the past three weeks that her exercise levels had been reduced.

Jane had always enjoyed school until comments were made about her body shape by some of her contemporaries. Mrs Smith said that, on reflection, the changes in Jane's behaviour and her marked obsession with dietary habits appeared to coincide with the time when these comments were made. Jane had changed from a lively girl to one who was withdrawn and sad.

Throughout the assessment, Jane continued to appear uninterested, giving the impression that the person being discussed was someone other than herself.

When Mrs Smith was interviewed on her own, she acknowledged that she had tried to seek medical assistance but Jane had refused to cooperate. Mrs Smith felt that she had been able to cope with Jane's behaviour until prior to the admission. At that point, Mrs Smith

became very concerned about Jane's welfare and, after some persuasion and bargaining, Jane agreed to consult her GP and be admitted to the ward.

Following the assessment, Jane's height and weight were measured and her weight was recorded as 39 kg.

Case background

Using the American Psychiatric Association's (1987) definition, anorexia nervosa is diagnosed using four criteria. These are:

1 the presenting weight is below 85 per cent that would be expected for an individual's weight and height

2 behaviours directed to maintaining this weight are present or reported

3 a phobia or dislike for the normal body weight and shape

4 amenorrhoea for at least three months

The diagnosis of anorexia nervosa is made difficult by the individual's denial of the existence of the problem and the conviction that they estimate their weight to be heavier than its real value. Individuals may also try to disguise their weight loss by wearing baggy clothing.

Anorexia nervosa may be induced by social/cultural, psychological, physiological and genetic factors (Raphael and Lacey 1992). It is more common in females than in males and the age of onset is commonly between 14 and 17 years. The onset of anorexia nervosa may be associated with certain physiological processes such as puberty and the associated changes in body shape (redistribution of body fat, development of breasts[1]) or other events such as leaving home, the divorce of parents, preparing for school examinations or being teased at school.

All referals are from the main reference at the end of this case study.
[1] Refer to pp.581–2 for a discussion of the onset of puberty.

There are two main methods by which an individual tries to achieve and maintain a weight that is below the expected norm and this helps to classify anorexia nervosa into its two types:

1   restricting type: during episodes of dieting, there is no binge-eating followed by either induced vomiting or misuse of laxatives

2   binge-eating/purging type: binge-eating and laxative abuse occurs during episodes of anorexia nervosa.

For cellular metabolic processes to occur, there has to be sufficient nutrient intake. Cellular activity requires energy and this is normally produced by the breakdown of glucose.[2] The amount of energy required is based on the basal metabolic rate and the physical activity level.[3]

The Department of Health also produces guidelines on the minimum eating requirements and this is promoted as part of a healthy eating campaign. According to these guidelines, Jane would require a dietary intake that would produce 2110 kcal a day. This means that Jane would have to consume approximately 500 grams of carbohydrate or 200 grams of fat a day to produce this amount of energy (respiration yields more than twice as much energy from fat than from carbohydrate, weight for weight). In anorexia nervosa, the intake of carbohydrates and fats is very much reduced, although the protein intake is usually adequate (Sharp and Freeman 1993).

It should be recognized that the estimated average requirements are guidelines only and some individuals may need more while others less and may be dependent on the amount of physical activity they take.

The other measure that can be used to estimate if an individual is consuming a normal diet is to compare that individual's diet with the mean-matched population weight. In this case, Jane's weight of 39 kg was much lower than that of the matched population,

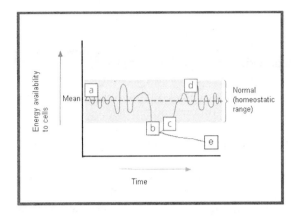

Figure 30.1   Anorexia nervosa. (a) Availability of energy within the range necessary for optimal cell function i.e. nutrient intake, digestion, delivery to tissues and assimilation are appropriate for health. (b) Reduced availability of energy. In anorexia nervosa, the problem is one of inadequate nutrient intake. (c) Therapies to increase energy availability. In anorexia nervosa, this would entail psychotherapeutic intervention and dietary advice . (d) Restoration of nutrient availability to within the range necessary for health. (e) Persistent malnourishment, as observed in chronic anorexia nervosa. Cell-functioning will gradually deteriorate and may even threaten the life of the individual.

being approximately 71 per cent of the expected weight.

When this nutrient intake falls, leading to a depletion in energy supply to the cells, an imbalance in homeostasis occurs (see Figure 30.1). During the initial stages of anorexia nervosa, the glycogen stores are used and when these are diminished fatty acids are used as the source of energy.[4] This would explain Jane's lack of subcutaneous fat.

As the episodes of enforced starvation progress in duration and severity, it is common to see the effects of this starvation on other body systems, including cardiovascular (bradycardia and hypotension), gastrointestinal (erosion of enamel and dentine from the surface of the teeth, oesophageal and intestinal erosion), renal (decreased glomerular

---

[2] Refer to pp.58–60 for a discussion of energy production by cellular respiration.
[3] Refer to p.106 for notes on basal metabolic rate (BMR).

[4] Refer to pp.59–60 for notes on the aerobic metabolism of fatty acids.

filtration rate resulting in an elevated serum urea and electrolyte imbalances), haematological (anaemia and thrombocytopenia) and endocrine (amenorrhoea) systems.

In addition to all these complications, the ability to regulate body temperature is also altered. The homeostatic mechanism of raising or lowering the core temperature in response to environmental changes is diminished and individuals who experience anorexia nervosa may require interventions to ensure that an ideal body temperature is maintained.

Dermatological changes are also noted. It is common to see fine downy hair, called lugano hair, especially on the face and on the back (Crisp 1980).

In addition to the physical effects, anorexia nervosa may also produce social and psychological symptoms. These may be exhibited as social isolation from friends, peers and family, restlessness in behaviour and irritability with others. There is a preoccupation with self rather than others and feelings of loss of control over one's life are also exhibited. This is an important element because of the close association between puberty and anorexia nervosa. It has been suggested that anorexia nervosa is the individual's response to puberty, which has brought about significant changes to their appearance. With starvation, it is hoped that these effects would be either stopped or reversed (Crisp 1980).

Care

As Jane's anorexia nervosa may have had a multifactorial cause, treatment must aim to address as many factors as possible. Treatment is based on three principles:

1 the restoration of body weight to recommence the normal physiological processes through dietary measures

2 the recognition of an altered perception of an individual's body, through individual therapy

3 the restoration of social interactions through group therapy

Although it was tempting to restore Jane's weight as quickly as possible, a certain degree of caution was necessary. Browning (1977) suggested caution because of three factors:

1 The individual with anorexia nervosa may have developed the ability to tolerate and cope with periods of starvation and electrolyte imbalances.

2 Any attempt at rapid weight gain may result in a degree of morbidity and even mortality.

3 The development of a long-term therapeutic relationship between the individual and health professional requires time at the beginning of the relationship for the development of rapport and trust (Browning 1977).

The dietary programme established for Jane was aimed at restoring her weight to that which is similar to the mean-matched population weight (see Figure 30.1). This involved a degree of negotiation between Jane, her named nurse and the dietician. The amount of food eaten at both meal times and in-between meals was recorded. Jane's weight was also measured and recorded daily.

Individual, group and family psychotherapy sessions were planned and Jane's initial involvement in these was minimal. She felt unable to recognize the severity of her condition and felt that it was the people around her who were experiencing the problem. But after the first two individual therapy sessions, Jane began to make some progress.

At the end of a fortnight of hospitalized treatment, Jane's weight had increased by 3 kgs. It was felt that she could be discharged home and the care could be continued on a day-patient basis.

Further information

Although at one time it was felt that anorexia nervosa was only present in Western and modern societies, there have been cases of anorexia nervosa reported in developing countries. Furthermore, recent media interest in anorexia nervosa may have given the impression that it is a modern-day phenomenon with an increasing incidence and prevalence. However, it is pertinent to point out that the term 'anorexia nervosa' was coined in 1874 and that any possible increase in incidence may be explained as an artefact in that there may be better recognition of the disease (Lucas *et al.* 1988). Anorexia nervosa, together with another common eating disorder, bulimia nervosa, remains fairly rare and the prevalence of these is approximately 1 per cent of adolescent girls (Crisp, Palmer and Calucy 1976).

Although this case study has only briefly examined some of the factors associated with the onset of the disease, there are many questions left unanswered. For example, the role of hunger or appetite has not been fully explored. It is recognized that the hunger and satiety centres situated in the hypothalamus have a role in the regulation of food intake.[5] In anorexia nervosa, there is some evidence that hunger is increased but the desire for the intake of food is suppressed.

Anorexia nervosa is a disease with a high degree of chronicity. It may last a lifetime and some individuals may die as a direct result. Although there are many modes of treatment of the disease, there is as yet no one method which is superior to others (Crisp *et al.* 1991).

Main reference

Clancy, J. and McVicar, A. J. 1995 *Physiology and anatomy: a homeostatic approach*. London, Edward Arnold.

Other references

American Psychiatric Association 1987 *Diagnostic and statistical manual of mental disorders* (DSM-II-R). Washington, DC, American Psychiatric Press.

Browning, C. H. 1977 Anorexia nervosa – complications of somatic therapy. *Comprehensive Psychiatry* 18, 399–403.

Crisp A. H. 1980 *Anorexia nervosa – Let me be*. London, Academic Press.

Crisp, A. H., Palmer, R. L. and Calucy, R. S. 1976 How common is anorexia nervosa? A prevalence study. *British Journal of Psychiatry* 78, 314–19.

Lucas, A. R., Beard, C. M., O'Fallon, M. W. and Kurland, L. T. 1988 Anorexia nervosa in Rochester, Minnesota: a 45-year study. *Mayo Clinic Proceedings* 63, 433–42.

Raphael, F. J. and Lacey, J. H. 1992 Sociocultural aspects of eating disorders. *Annals of Medicine* 24, 293–6.

Sharp, C. W. and Freeman, C. P. L. 1993 The medical complications of anorexia nervosa. *British Journal of Psychiatry* 162, 452–62.

[5] Refer to pp.121–2 for a discussion of the hypothalamic role in the regulation of food intake.

31 The case of a young woman with occupational hyperstress

Judith Tyler

Learning objectives

1 To revise the concept of stress from an historical physio-psychological perspective to a developing awareness of this in terms of homeostatic balance

2 To understand the interrelationship between different causative factors and effects in relation to stress

3 To appreciate individuality in establishing the stress thresholds within which a homeostatic balance can be achieved between capabilities and demands

4 To understand the concept of occupational stress

5 To be aware of the role of a healthy stress response in restoring the homeostatic balance in a person experiencing occupational hyperstress

6 To understand the rationale for care and support of a woman experiencing occupational hyperstress

Case presentation

Polly is a 23-year-old paediatric nurse. Previously enthusiastic, sensitive and compassionate, during the past six months Polly has become cool and detached towards her patients, uncommunicative with their parents and aloof and distant from her colleagues. Her actual delivery of care has remained competent, but she feels beset by doubts about her own ability, constantly checking and rechecking procedures and equipment with which she has been involved. There never seems time to complete everything expected of her and the increasing difficulty which she finds in expressing herself in the nursing-care plans means that this aspect of her work is left to the last possible moment. She has begun to dread returning to work following her off-duty and has missed several days suffering from migraine.

Polly lives with her boyfriend, Tom, who has little patience with her current moods. He feels that if she no longer likes children she should change her job.

Case background

This case relates to an individual who is currently experiencing a condition of hyper-stress thought to be principally occupational in origin. The study of stress has been variously approached from a predominantly physiologi-cal or from a psychological and social environ-mental perspective, with Thompson (1982) adding the developmental dimension. This reflects a common and compelling research interest founded in formerly distinct academic disciplines. Current thinking favours a more eclectic approach, and one which recognizes the complex interrelationship between mind and body, cause and effect. Stress may be viewed, therefore, as a psycho-physiological homeostatic imbalance which arises when there is an actual or perceived demand–capability mismatch between the individual and their environment.

Although the concept of stress is relatively modern, its study has attracted an immense amount of interest, and a variety of definitions and models have emerged.[1]

Stress has been identified as causing (as a stimulus) or as constituting a response to poten-tially damaging internal or external demands on the individual. Earlier stimulus-based models introduced a definition of stress which has persisted to the present day and which antici-pates the current concept in terms of homeo-static balance. Here stress is seen as 'any internal or external stimulus which disturbs the equilibrium of the body' (Thompson 1982).

This focus produced a preoccupation with the causative components of stress which was to dominate research in this field. Even Selye (1956), who pioneered the response-based model of stress, explored various environ-mental stressors (causes) in relation to the

physiological changes which occur in response. Subsequent work has developed the concept of negative stressors, and sought to identify these in relation to significant life events (Holmes and Rahe 1967).

Early research explored the relationship between perceived occupational stress and job satisfaction in a variety of white-collar, managerial, professional and semi-skilled employment. This pattern has applied to the study of stress in nursing, with continuing concentration on the causative elements of negative stress in a variety of different special-ities. The results of negative stress have been seen in terms of occupational dissatisfaction, disengagement and 'burn out'.

While the traditional approach has concen-trated on the negative effects of stress,[2] the contemporary stance, in defining stress in terms of homeostatic balance, admits to the concept of deleterious and beneficial stress: respectively 'distress' and 'eustress'.[3] Furthermore, a normal individual range is established, beyond the bounds (stress thresholds) of which that person can be said to be experiencing hyperstress or hypostress (see Figure 31.1).

Care

In assessing Polly's need for support and pos-sible guidance in her current state of homeo-static failure, an important early measure must be to get her to attempt to distinguish between the actual causes of her distress, always remembering that all aspects of an individual's life and the responses they make to the various situations in which they find themselves are interrelated.[4] It seems probable that the princi-pal cause of Polly's hyperstress is occupation-ally based, as she is showing the characteristic symptoms of 'burn out': an extreme reaction to

---

All referals are from the main reference at the end of this case study.
[1] Refer to pp.637-9 for definitions of stress.

[2] Refer to Table 22.1 for a discussion of the effects of stress.
[3] Refer to pp.639–40 for a definition of eustress and distress (see also Table 22.2).
[4] Refer to pp.642–5 for a discussion of the subjectivity of stressors.

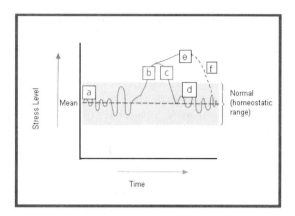

Figure 31.1 Occupational hyperstress. (a) Stress level maintained within homeostatic norms (perceived capabilities = perceived demands). (b) Distress; stress threshold superseded: Hyperstress (perceived capabilities < perceived demands). Stress-related signs and symptoms may be observed (e.g. migraine, heartburn, etc.). (c) Reduction of stress by identification of stressors and stress management (capability–demand mismatch is reduced). (d) Stress level returns to homeostatic norm. (e) Failure of individual coping mechanisms. Exacerbation of stress-related signs and symptoms leading to stress-related disorders. (f) Intervention: provision of health-care support to control signs and symptoms and to facilitate coping.

unrelieved stress which has long been associated with those engaged in the 'caring professions' (Menzies 1960).

These symptoms can be summarized as disaffection and withdrawal of involvement with those for whom they care. However, Tom's recent lack of sympathy, while it may be purely a response to the change he sees in Polly, could be indicative of deeper problems within the relationship which may have impacted on Polly's ability to cope with the demands of her job.

Appropriate support for Polly will only be identified once the predominant source of her distress is recognized. As Polly's stress is thought to be occupationally founded, support must be provided in the workplace. This will operate at two levels: with individual support for Polly through the involvement of the occupational health department, and through

the awareness of her manager, but it should include an assessment of the current environment within the paediatric unit where she works. It may be that Polly's stress is not isolated but indicative of a general job dissatisfaction among her colleagues. Any recent changes in practice (such as altered shift patterns) and any undue pressures should be identified and, where possible, measures taken to address them.

Individual support for Polly should focus on the following areas:

- Those caring for her, and Polly herself, must recognize that the level of stress she is experiencing is outside the usual limits (thresholds) within which her previous homeostatic norm has naturally fluctuated. During this time, the demands made on her were balanced by her individual capability (see Figure 13.1).
- Care for Polly will need to restore the homeostatic balance by reducing stress levels to below the upper threshold. Specialist psychological intervention may be indicated, with the appropriate medical treatment for her presenting physiological problems, such as migraine.
- An individual programme of stress reduction should be instituted, incorporating techniques whereby Polly will be able to recognize and use the beneficial effects of everyday stressors (eustressors) to counteract the adverse effect of her current hyperstress.
- Medical, nursing or specialist-counselling intervention will need to address the occupational focus of her distress as follows:
- Career guidance: has her original ambition in joining the paediatric unit been realized?

IF SO

- Does she still feel that this is appropriate?
- How can she be developed in the role?
- What will facilitate her career development? e.g. continuing education, enhanced support through clinical supervision.

IF NOT

1   What alternatives are there?

2   How may she access them? e.g. possibly leaving the speciality or even leaving nursing.

In either eventuality, it is important to concentrate on Polly's strengths and aspects about which she feels positive and less distressed.

The importance of a non-judgemental approach is paramount. Hyperstress and burnout is known to affect those who are caring and conscientious. Above all, a level of care should be instituted which effectively restores the state of balance between demands and capabilities, irrespective of Polly's ultimate decisions about her work and her relationships.

Further information

There is a comprehensive body of research into stress, burnout and coping mechanisms in nursing. Acute and technological areas of care have long been associated with hyperstress in nursing.

In a study of women in employment, Fletcher, Jones and McGregor (1991), found clinically significant levels of stress to be higher among nurses than among those in a general population sample.

Note that in the context of assisting an individual/individuals to return to a level of stress within the normal homeostatic range defined by their stress thresholds, the model of clinical supervision which is advocated is one of support and development, rather than the more narrow regulation and standard setting.

Main reference

Clancy, J. and McVicar, A.J. 1995 *Physiology and anatomy: a homeostatic approach.* London, Edward Arnold.

Other references

Fletcher, B., Jones, F. and McGregor, C. J. 1991 The stresses and strains of health visiting: demands, supports, constraints and psychological health. *Journal of Advanced Nursing* 16, 1078–89.

Holmes, T. H. and Rahe, R. H. 1967 The social readjustment scale. *Journal of Psychosomatic Research* 11(2), 213–18.

Menzies, I. E. P. 1960 Nurses under stress. *International Nursing Review* 7, 9–16.

Selye, H. 1956 The general adaptation syndrome and diseases of adaptation. *Journal of Clinical Endocrinology* 6, 117–18.

Thompson, R. 1982 A pocket guide to stress. London, Arlington Books.

32 The case of a man with occupational hypostress

Judith Tyler

Learning objectives

1 To revise the concept of stress from an historical physio-psychological perspective to a developing awareness of this in terms of homeostatic balance

2 To understand the interrelationship between different causative factors and effects in relation to stress

3 To appreciate individuality in establishing the stress thresholds within which a homeostatic balance can be achieved between capabilities and demands

4 To understand the concept of occupational stress

5 To acknowledge the concept that hypostress represents a deviation from the normal state of homeostatic balance, and that as such it is potentially injurious to health

6 To be aware of the role of a healthy stress response in restoring homeostatic balance in a person with occupational hypostress

7 To understand the rationale for care and support of a man experiencing occupational hypostress

Case presentation

Gordon is 43 years old and has recently lost his job as manager of a farm when the owner sold it to a large cooperative with its own management structure. He has been applying for similar positions without success: a factor he attributes to his age and to his reluctance to move out of the area and disrupt his children's secondary schooling. He has had to leave the tied house which went with this post, and move into rented council accommodation. To continue to support his family, Gordon has reluctantly accepted a position as tractor-driver on a neighbouring farm.

The reality of the situation is that Gordon has moved from a position of considerable responsibility and great variety, managing a team of men and a substantial budget, on a large farm with mixed arable enterprises and a pedigree breeding stock of pigs, sheep and cattle. His current role is subordinate; his salary, status and standard of living is reduced, and he finds the work repetitive, boring and lacking in any stimulation or challenge.

Gordon sees his GP, complaining of tiredness, poor sleep and indigestion.

Case background

The case background of the young woman with hyperstress should be read in conjunction with this. This one outlines the development of research into the nature of stress, the manifestation of occupational stress, and the homeostatic imbalance which can occur in situations of hypostress.

A certain amount of stress is necessary for survival. At a simplistic level, stressors may be seen as 'prompts' for basic functions to occur. Where these cease to exist, the organism will die.

Normal fluctuations in levels of stress are experienced universally. Where these remain within the boundaries of the stress thresholds shown below, homeostatic balance is achieved between the demands placed on an individual and that person's ability to cope.

To date, research has concentrated on hyperstress, with only a comparatively recent recognition of the potentially, equally injurious, situation where sustained hypostress occurs (Maslach and Jackson 1981). Where significant hypostress develops, the resulting homeostatic deficit can lead to increased psychological and physiological morbidity.[1] The common symptoms of hypostress include boredom, lethargy and chronic tiredness. There may be disaffection, a lack of motivation and a range of psychosomatic disorders occasioning an increased incidence of sickness and absenteeism. Severe depression may arise, with the risk of suicide.

All referals are from the main reference at the end of this case study.
[1] Refer to p.640 and to Table 22.1 for a discussion on morbidity arising from distress.

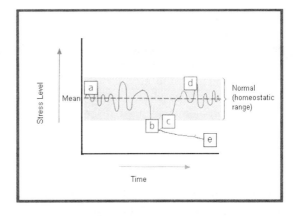

Figure 32.1 Occupational hypostress. (a) Stress levels maintained within homeostatic norms (perceived capabilities = perceived demands). (b) Distress; hypostress (perceived capabilities > perceived demands). Signs and symptoms of hypostress will occur (e.g. poor sleep, indigestion, tiredness, etc.). (c) Use of alternative stimuli to promote eustress. This may be prompted by clinical intervention (d) Restoration of stress levels to homeostasis norms. (e) Failure to restore capability: demand mismatch. Prolonged hypostress, with exacerbation of signs and symptoms leading to stress-related illness (e.g. depression).

Care

Given the situation as it is presented, Gordon is probably suffering from occupational hypostress. His description of boredom and frustration in his work, and the feeling of 'being tired doing nothing' support this. However, he has recently experienced significant trauma in losing a responsible job and an established home, and these events themselves cause distress. This emphasizes the difficulty in determining a rigid 'diagnosis' of hyper or hypostress, and endorses the concept of constantly fluctuating levels of stress (see Figure 31.1 and 32.1).

In describing Gordon's care, the assumption has been made that his earlier stress response to his personal trauma has to some extent been resolved, and that the current manifestation is primarily caused by hypostress in his diminished occupational role.

Gordon's care will require the appropriate investigation and symptomatic treatment of his presenting ill-health. His GP will wish to exclude other physiological causes for his tiredness (e.g. possible anaemia), and establish the reason for his indigestion (e.g. ulcer, hiatus hernia, etc.). The sleep problem will need to be thoroughly investigated, as a history of disturbed rest and early morning waking may indicate a level of clinical depression which will require immediate intervention to prevent suicidal behaviour.

While these investigations are being conducted and the treatment, and where necessary, specialist referral is being undertaken, Gordon will need help to restore his presenting homeostatic imbalance (see Figure 32.1).

Gordon's current level of stress has fallen beyond the lower threshold of the established homeostatic range, and is presenting as distress. This has caused stress-related symptoms and current medical intervention is addressing these. The psychological care he needs will aim to restore homeostasis by helping him to cope with the negative 'unhealthy' stress response he is experiencing.

Gordon needs to be able to utilize everyday beneficial stressors (eustressors[2]) which will help to restore this balance. He has suffered an inevitable loss of self-esteem, and the lack of challenge and respect in his current work endorses this. He may be experiencing a sense of shame where his family and wider circle of friends are concerned, tension in his relationships and feelings of profound isolation.

Measures to promote the beneficial effects of eustressors which will provide an alternative source of stimulation and a necessary 'healthy' stress response might include working with Gordon to get him to recognize the remaining areas of challenge in his life.

The fact that Gordon has sought medical opinion is encouraging as it shows a recognition on his part of the need for action to restore his equilibrium. Support for Gordon will need to capitalize on and develop this self-awareness, to promote alternative strategies for him to adopt in relation to his work. It may be that he can be encouraged to find some unrealized potential in the job or he may be able to find this fulfilment in other interests, family, hobbies or organizations. Given his former level of responsibility and involvement, the challenges he needs to develop to establish a pattern of healthy stressors in his life are more likely to be found in social/community activities where he can use his former expertise and regain his self-respect.

By developing Gordon's potential to recognize and use the beneficial effects of stress, a return to normal homeostatic function can be achieved.

Further information

The subject of occupational hypostress is a developing area of stress-related research. Studies which have found this to be a factor in unhealthy stress responses among nurses include Cherniss (1980), and Kelly and Cross (1985). Occupational areas which make low demands on the workers but provide little support (or actual constraint) in fulfilling those demands are potentially more stressful than highly demanding areas where there is a good level of support (Maloney 1982).

The concept of 'personality hardiness' has evolved, showing that where an individual can exercise some control in their working situation, they are less likely to experience hypostress (Kobasa, Maddi and Kahn 1982).

Main reference

Clancy, J. and McVicar, A. J. 1995 *Physiology and anatomy: a homeostatic approach*. London, Edward Arnold.

Other references

Cherniss, C. 1980 *Professional burnout in human service organisations*. New York, Praeger Scientific.

---

[2] Refer to pp.639–40 for a definition of eustress (see also Table 22.2).

Kelly, J. G. and Cross, D. G. 1985 Stress, coping behaviours and recommendations for intensive care and medical surgical ward registered nurses. Cited in *Research in nursing and health* 8, 321–8, Chichester, John Wiley.

Kobasa, S. C., Maddi, S. R. and Kahn, S. 1982 Hardiness and health: a prospective study. *Journal of Personality and Social Psychology* 42 (1), 168–71.

Maloney, J. P. 1982 Job stress and its consequences on a group of intensive care nurses. *Advances in Nursing Science* January, 31–42.

Maslach, C. and Jackson, S. E. 1981 The measurement of experienced burnout. *Journal of Occupational Behaviour* 2, 99–113.

33 The case of a young man with schizophrenia

Joan Schulz

> *Learning objectives*
>
> 1 To identify the possible causes of schizophrenia
>
> 2 To explore the biopsychosocial basis of schizophrenia
>
> 3 To appreciate some interventions which may facilitate improvement for the client

Case presentation

Paul is a 17-year-old man who is single and lives at home with his mother. He is employed as a storeman in a large furniture warehouse. Until recently Paul had no previous significant medical history, although he has been described as quiet.

On her return from holiday Paul's mother found him to be withdrawn, uncommunicative, anxious and exhibiting what she described as 'bizarre behaviour'. He believed he was receiving messages from the television set and his favourite records. Over a period of two to three weeks, he became increasingly difficult to manage, refusing to eat his meals as he believed they were poisoned. He would stay awake for long periods during the night and occasionally would be verbally aggressive towards his mother. He refused to leave the house and attend work, and eventually he shared his fear with her that the house was being watched and 'they' were spying on him. At this point his mother contacted their GP.

Case background

Schizophrenia is a group of disorders manifested by characteristic disturbances of mood and behaviour. There are many diverse research explanations as to the origins of schizophrenia. There are some indications that biological disposition to schizophrenia is caused by an excess of the neurotransmitter dopamine within the hippocampus of the brain.[1] This may be as a result of producing abnormal dopamine molecules, or from producing the dopamine in excessive amounts (Iverson and Iverson 1975).

There are also some indications that there may be a disposition to schizophrenia being polygenically inherited (Kendler and Diehl 1993).[2] Some research has provided strong evidence that there are greater risks of developing schizophrenia the

All referals are from the main reference at the end of this case study.
[1] Refer to p.372, and pp.365–9 for a description of the hippocampus and a general discussion of the role of neurotransmitters.
[2] Refer to pp.604–5 for a discussion of polygenic inheritance.

closer the blood relationship is to the sufferer (Tsuang 1982), but there is insufficient evidence to state that schizophrenia is unequivocally a genetic disorder.

Psychological and social perspectives on schizophrenia have become much more popular in schizophrenia research over the last three decades. These theories offer a greater recognition of the multifactorial models which seek to explore the interactions of biological, social and psychological factors influencing the potential physical and mental health problems for the sufferer. Laing (1967) has written prolifically about his view on schizophrenia and suggests that it may consist of 'normal responses to an abnormal situation; this being the problems encountered in faulty family dynamics'.

Vulnerability models such as those discussed by Zubin and Spring and Nuechterlein and Dawson (cited in Birchwood and Tarnier 1995) help the practitioner to understand the interactions between the environment and the individual's responses to these factors.

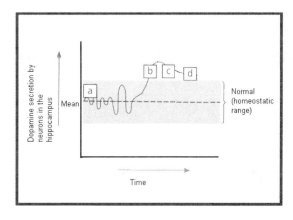

Figure 33.1 Schizophrenia. (a) Secretion of the neurotransmitter within dopaminergic pathways appropriate for 'normal' emotional and behavioural responses. (b) Hypersecretion of dopamine leading to an imbalance of activity of the neurological pathways. Cognitive symptoms and schizophrenia are observed. The causes and levels of hypersecretion may be indicative of the severity of signs and symptoms. (c) Reduced dopamine secretion. In health this would be a physiological response which helps to maintain *cognitus*-processing within norms. (negative feedback response). In schizophrenia there seems to be a failure of such processes, requiring medication and cognitive therapies to facilitate the change. (d) Incomplete return to homeostatic range. Symptoms will persist but with less severity.

Care

Medical intervention was required in order to correct any neurochemical imbalance (see Figure 33.1) and Paul was commenced on the neuroleptic drug chlorpromazine. This is often used as it can help with anxiety symptoms and the behavioural disturbances which Paul sometimes exhibited. Paul was experiencing distressing symptoms which affected his normal cognitive functioning. The by-product of chlorpromazine is that it also has a sedative effect which helped to reduce Paul's anxiety and to improve his sleep pattern. A community psychiatric nurse visited Paul, initially twice a week to monitor the effect of the medication, and to initiate and develop a therapeutic relationship with both Paul and his mother. These visits were subsequently reduced to once a week.

Paul was encouraged to express his feelings initially to the nurse and then in joint sessions with his mother. Relatives often do not understand the problems their loved ones are experiencing and at times may be intolerant of the sufferer's behaviour. The nurse would explore the family dynamics using a pragmatic and flexible approach.

A cognitive behavioural model would be adopted to encourage Paul to change his thinking pattern and to increase his self-esteem. In this way the 'bizarre behaviour' observed by his mother should lessen.

The goals negotiated and agreed with Paul would be:

1   To ensure an adequate dietary intake

2   To have some structure added into his day to avoid sleeping or withdrawing. Various options were suggested. Paul chose to join a

local group run as a social support by the community psychiatric nurse and volunteers

3 To gradually socialize with his close friends and take up his previous leisure pursuits

4 To remain on current medication and be monitored by a CPN and a GP on a regular basis

Further information

The symptoms of schizophrenia include delusions (false beliefs), and hallucinations (false perceptions), and these can take the form of hearing voices, and visualizing, as the most common types. There are also difficulties in thinking, feeling and behaviour.

Environmental factors can affect the person and stress and tension make the symptoms worse, possibly triggering exacerbation of the illness. Such factors range from problems at birth, peer-group problems, social stressors, relationships within the family (high expressed emotion) or any other life event which may trigger the onset of the illness (Leff 1976; Lewis *et al.* 1981).

Main reference

Clancy, J. and McVicar, A. J. 1995 *Physiology and anatomy: a homeostatic approach*. London, Edward Arnold.

Other references

Birchwood, M. and Tarnier, N. 1995 *Innovations in the psychological management of schizophrenia*. Chichester, John Wiley.

Iverson S. D. and Iverson, L. L. 1975 *Behavioural pharmacology*. New York, Oxford University Press.

Kendler, K. S. and Diehl, S. R. 1993 The genetics of schizophrenia: a current genetic-epidemiological perspective. *Schizophrenia Bulletin* 19 (2), 261–85.

Laing, R. D. 1967 *The politics of experience*. New York, Pantheon.

Leff, J. 1976 Schizophrenia and sensitivity to the family environment. *Schizophrenia Bulletin* 2, 566–74.

Lewis, J., Rodnick, E. H. and Goldstein, M. J. 1981 Intrafamilial interactive behaviour, parental communication deviance and risk for schizophrenia. *Journal of Abnormal Psychology* 90 (5) 448–57.

Tsuang, M. T. 1982 *Schizophrenia. The facts*. Oxford, Oxford University Press.

The case of a man with Alzheimer's disease

34

Roy Bishop

Learning objectives

1 To understand the impact of disease-led brain dysfunction on the individual and those around him

2 To recognize the progressive nature of Alzheimer's disease and the problems this presents for attempts to maintain the person's quality of life

3 To appreciate the importance of support systems in maintaining homeostatic equilibrium in the relationships between the patient and his carers

Case presentation

Mr Harris is aged 68. He is a retired schoolteacher who is married with a wife, three children and two grandsons. Over the past two years his memory for recent events has deteriorated and his previous, rather pleasant, outlook on life has changed to that of an irritable and sometimes aggressive person. Following a disagreement with his wife about his judgement in handling the family finances, he became very upset and stormed out of the house and into the path of a passing car. Fortunately Mr Harris suffered only a few minor bruises but staff at the accident unit were concerned enough about his mental state to keep him in hospital overnight.

The following morning Mr Harris was noticed to be rather more confused than might be expected in the circumstances. When his wife described the changes she had seen in him over the past two years, medical staff began to suspect that, in the absence of any head injury, a disease process affecting brain function might lie behind his altered behaviour. The history given by Mrs Harris pointed quite firmly toward the presence of a developing form of dementia known as Alzheimer's disease.

Case background

Alzheimer's disease is a form of dementia which impairs the higher mental functions such as memory, problem-solving, judgement and emotional reactions (see Figure 34.1). This disturbance occurs without clouding the consciousness of the individual and follows a progressive course which can continue for a period of years.

The memory disturbance is usually for recent events and is very often the most noticeable feature in the early stages of the disease.

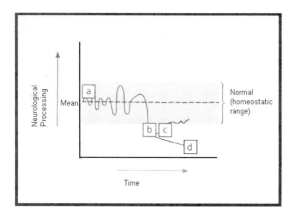

Figure 34.1 Alzheimer's disease. (a) Balance between neurology pathways involved in neural processing/appropriate for normal cognitive functioning. (b) Imbalance of neural pathways arising from brain tumour, CVA or slow onset loss of brain cells (e.g. ageing, Alzheimer's disease). (c) Partial restoration of neural functioning by reinforcement therapy, or (in the case of CVA) improvement of cerebral circulation, or by using drugs to modify neurochemistry. (d) Persistent deterioration in neural functioning observed in Alzheimer's disease as a result of a progressive loss of neurones.

A further, and related, problem is an associated deterioration in language ability.

Common pathological changes in Alzheimer's disease are the loss of neurones (brain cells) in certain areas of the brain, neurofibrillary tangles (abnormal changes in the cytoplasm of brain cells) and neuritic plaques (clusters of degenerating nerve endings).[1,2] In recent years a number of theories related to chemical changes have emphasized the involvement of noradrenaline and serotonin,[3] as well as a reduction in cholinergic activity in the cerebral cortex (Thompson and Mathias 1994). The cause of these pathological changes is as yet unknown, but theories of inheritance and of environmental influence are under investigation. Unfortunately the disease is very difficult to diagnose with certainty

All referals are from the main reference at the end of this case study.

[1] Refer to pp.381–3 for notes on the parts of the brain involved in cognitive processes.

[2] Refer to pp.351–3 for a description of nerve-cell structure.

[3] Refer to pp.365–7 for a discussion of neurotransmission.

during life, so much evidence has to be gained post-mortem.

Care

Mr Harris was assessed by physicians, psychiatrists and psychologists in order to determine the level of malfunction and the management of his disease. The care of people with Alzheimer's disease is mainly aimed at compensating for and limiting the negative consequences of dementia (Wright and Giddey 1993). Nurses, occupational therapists and physiotherapists will have a very significant part to play in the professional care of Mr Harris but a large burden may also be placed on his family and their relationships with him and with one another.

Drug treatment is, as yet, relatively ineffective in either halting or improving the disease process. However, some drugs such as arecoline and 4-aminopyridine have been observed to help in redressing the chemical imbalance associated with the disease. Other drugs may be used to lessen the effects of confusion so commonly seen in Alzheimer's disease.

Specific nursing care involves such activities as reality orientation, risk assessment and protection and help with maintaining the activities of living while encouraging optimum independence.

Because the disease is progressive the care of Mr Harris over what may be some years may take a toll on his family and while health-care professionals can ease this pressure by providing caring and sensitive support, they can never compensate for the loss of the husband and father they all knew. Families can easily become dysfunctional because of these pressures and the professional carers must work closely with them in order to help them maintain equilibrium. Mr Harris is likely to develop more disturbing and unpleasant symptoms and, while medical and nursing care can relieve much of his discomfort, his family will need to know that help is also available for them.

From a more positive perspective it is vital to help Mrs Harris feel a valued and significant contributor to her husband's care. As Mr Harris's disease is at a relatively early stage, Mrs Harris is likely to find there is much she can do to help him maintain as much of his normal functioning as possible. In the past, early hospitalization has been shown to lead to problems of institutionalism and more rapid decline, where professionally supported home care can help sufferers and relatives maintain comparative normality for significantly longer. Effective homeostasis in Mr Harris's case can be encouraged by ensuring the availability of professional care and the on-going assessment which is supportive of the loving input of his own family.

Further information

Alzheimer's disease was first described by a German physician, Alois Alzheimer, and the condition was named after him. It was originally attributed to women in their early fifties and was called presenile dementia.

Alzheimer's disease usually leads to death in approximately eight to ten years, although it can progress much more quickly (three to four years) or more slowly (as much as 16 years). Occasionally the disease progresses slowly for years and then more rapidly. The relative stable periods are referred to as 'plateaux'.

Main reference

Clancy, J., and McVicar, A. J. 1995 *Physiology and anatomy: a homeostatic approach*. London, Edward Arnold.

Other references

Thompson, T. and Mathias, P. (eds) 1994 *Lyttles mental health and disorder*, 2nd edn. London, Baillière Tindall.

Wright, H. and Giddey, M. (eds) 1993 *Mental health nursing – from first principles to professional practice*. London, Chapman & Hall.

35

The case of a man with depression

Derek Shirtliffe

Learning objectives

1 To identify the possible causes of depression

2 To understand some of the underlying physiological basis of depression

3 To appreciate the approaches which inform the treatment of depression

Case presentation

William is a 30-year-old man who is married with one child who is five years of age. He has worked for 12 years as a miner in a colliery near to his home. He has now been made redundant and after seeking employment for six months he has not found an alternative job. He received redundancy pay but the amount is small in comparison with the outstanding mortgage on his home.

William's wife has a part-time job and feels that she would like to expand the number of hours she works for both financial and social reasons. William is resistant to this development: he feels he should provide for his family and this emphasizes his feelings of inadequacy and worthlessness. Their present circumstances have been the source of many arguments between William and his wife. Each is becoming increasingly distanced by the frustration they feel at the intransigence of their spouse. William has loss interest in his daughter with whom he previously interacted and played regularly. He is described by family friends as withdrawn and lethargic. He

eats without enthusiasm and has lost weight: this is remarked upon by people who know him well. He previously liked to read extensively but he now finds it difficult to concentrate beyond short periods of time. He suffers from insomnia and feels tired and fatigued during most of his waking hours.

Case background

The term depression is frequently used in everyday language to describe normal downswings in mood and thus many people in any population are affected at a given moment in time. These downswings are normally of short duration and may serve to highlight a specific loss and activate the person to establish links with people who may be able to help. However, at other times the depression may deepen to such an extent that it serves to attract further symptoms which are more problematic than helpful.

A physiological disposition to depression appears to be associated with a deficiency of the excitatory transmitters noradrenaline and serotonin in certain parts of the brain, leading to excessive expression of inhibitory pathways.[1]

Higher centres of the brain such as the hypothalamus and the cerebral cortex can modify arterial blood pressure via the medullary centre of the brainstem in response to strong emotions. So during depressive episodes there can be a decrease in vasomotor sympathetic reflex with a resultant decrease in blood pressure.[2]

The psychological and sociological perspectives of depression have spanned a range of possible formulations which influence the development of the condition. Some theories which have been proposed as leaving people vulnerable to and maintaining the depressive state have been behavioural in emphasis, involving notions such as reduced social reinforcement, the inability of the person to reinforce their own behaviour and learned helplessness. Some cognitive theories of depression have embraced three main ideas. First, the concept of negative thoughts spontaneously arising in the individual. Second, the component of systematic logical errors in the thinking of depressed individuals. Third, the notion of the presence of depressogenic schemas, which are long-standing attitudes or assumptions about the world which individuals use to organize their past and present experience (Beck 1976).

Care

William is experiencing a major depressive episode. The depression is severe, without psychotic symptoms, but it has markedly interfered with William's usual social activities and relationships. The main emphasis of the care adopts a cognitive–behavioural approach. The underlying assumption which supports this

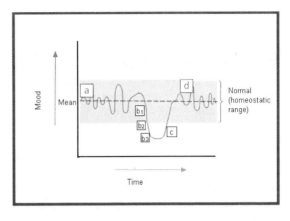

Figure 35.1 Depression. (a) Emotional state within normal 'mood swings'. (b_1) Genetic predisposition to depression tending to move mood state toward lower end of homeostatic range. (b_2) Influence of life events to exacerbate change in mood. (b_3) Imbalance in excitatory neurotransmitters disrupts neurological functions resulting in depressive illness. (c) Use of tricyclic antidepressant drugs to restore neurotransmitter balance, and influence of cognitive therapies to promote neurological functioning. (d) Return of emotional state to normal range.

strategy is that the client's depression follows from distortions in thinking. Treatment therefore seeks to change the patterns of thinking (see Figure 35.1) and bring about a long-lasting resolution of the problems.

Some general characteristics of cognitive therapy give structure to the approach (Beck 1976). First, the therapist reviews with the client what subjects are to be discussed during the session. Second, the therapist structures the therapy time so that a balance is achieved between the central and peripheral issues. Third, the therapist periodically summarizes during the session and invites the client's reaction to the summary. Fourth, the session is dominated by a questioning approach by the therapist. Statements of fact or offerings of advice are avoided. Finally the therapist agrees with the client that they will reflect in time outside the sessions on areas which proved problematic. The client is also asked to sum up what has happened during the session and to indicate what they feel has been helpful, inappropriate or hurtful about it. The client is

All referals are from the main reference at the end of this case study.
[1] Refer to pp.366–9 for a description of inhibitory and excitatory synapses.
[2] Refer to pp.236–8 for a discussion of blood pressure control reflexes.

also asked to explain what they think is required of them during the further reflection time.

The restoration of homeostasis can be facilitated by the use of tricyclic antidepressants which prevent the reuptake of noradrenaline from the synaptic cleft[3] and prolong the activation of post-synaptic receptors. This redresses the imbalance of neurotransmitters observed in the condition.

William also has begun a course of these drugs and nurses are monitoring his response in order to respond to their side-effects. Alongside the use of medication, nurses working with William have encouraged him to articulate the emotional pain he is experiencing following the loss of his job in both individual and group situations. He has been encouraged to discuss his feelings with his wife and daughter. This promotes increased communication between husband and wife who have become increasingly estranged from each other. It therefore serves the function of enabling William to increase his feelings of support and decrease his feelings of isolation.

Another objective of nursing intervention is to help William to regain a role similar to that implicit in his former employment. With this in mind, he has recently begun a programme of preparation for job-seekers, involving application and interview skills. One aspect of this intervention is that William has made some friends among other people on the programme. This has also helped to increase his social support and has provided an impetus for him to undertake further study for an alternative career.

William's difficulty in sleeping during the night is also being addressed: he is being asked to record and report back about his sleep patterns. He has agreed that he will have no drinks containing caffeine, refrain from watching television and indulge in little or no conversation for half-an-hour before he goes to bed. This regime is intended to facilitate the reduction of stimulants and stimulation,

thereby helping the return of predepression sleep patterns.[4] The nurses involved in William's care are monitoring his diet carefully so that his lack of interest in food is not impairing his wellbeing.

Further information

Generally, depression is more prevalent among women than men and evidence implies that this may be because more women than men experience a relapse. Controversy surrounds the long-standing idea that depression can be ascribed to 'endogenous' or reactive influences. Originally this distinction attempted to differentiate between people whose depression arose from a biochemical interaction of the brain and those who were depressed in reaction to external stresses.[5]

More recently, the term endogenous has been used to describe a number of symptoms and not to how the depression was caused. Depressive illness appears to be caused by a combination of physiological, psychological and social factors. What remains unclear is whether biochemical changes cause alterations of mood or are the effect of alterations of mood.

Main reference

Clancy, J. and McVicar, A .J. 1995 *Physiology and anatomy: a homeostatic approach*. London, Edward Arnold.

Other reference

Beck, A. T. 1976 *Cognitive therapy and the emotional disorders*. New York, International Universities Press.

[4] Refer to pp.380–1 and pp.671–2 for a discussion of sleep, and the circadian implications of the effects of sleep respectively.
[5] Refer to pp.638–40 for a discussion of the effects of stressors.

[3] Refer to pp.365–6 for a discussion of synaptic functions.

The case of a woman with phobic anxiety

36

Roy Bishop

Learning objectives

1 To understand phobic anxiety as an exaggeration of normal psychobiosocial functioning

2 To appreciate the intensity of emotion and the subjective experience of phobic anxiety

3 To understand the rationale behind therapeutic intervention in phobic anxiety

Case presentation

Cathy first consulted her GP complaining of feelings of unease which seemed to be worse when she was in the company of other people. She felt that at any moment she would embarrass herself by fainting or just falling over. Often her heart would pound so hard she felt that everyone could hear it and she thought that she might be about to have a heart attack. She asked her GP if he would check her heart because she was convinced that she must have a defect of some sort.

At the surgery the GP examined Cathy and performed an electrocardiograph (ECG)[1] and found no indication of disorder to support Cathy's fears. At first she was reassured but she soon began to wonder if the doctor might have made a mistake and she returned to ask him to examine her again. This alerted the

doctor to the possibility that Cathy was suffering from an anxiety-related disorder.[2]

On further examination Cathy revealed that her feelings of anxiety and fear of an impending heart attack seemed to be much worse when she was in the company of groups of people. At first she managed to cope with this but the feelings became more and more intense and she found herself avoiding groups because the thought of being with them made her feel so frightened. She felt she had now reached the stage where she could not face meeting people and had even begun to avoid groups of people she knew very well. The doctor noticed that Cathy became very tense and looked in obvious distress[3] even when just talking about a situation which usually causes most people

---

All referals are from the main reference at the end of this case study.
[1] Refer to pp.212–13 for notes on the electrocardiogram.

[2] Refer to p.619 for brief notes on anxiety association with pain, p.625 for its connection with personality and myocardial infarction, and pp.655–6 for its association with pseudo-organic and organic disease.
[3] Refer to pp.647–52 for a discussion of the subjectivity of the stress response.

little anxiety. As the interview progressed Cathy revealed more and more of her fear that she would make a fool of herself in front of others and it was clear that she would need help to control this fear which was threatening to dominate every aspect of her life.

Case background

Anxiety is a normal human emotional and physical response and placed in threatening situations it is usual to respond in such a way that we are able to face the threat or avoid it causing us harm. The perception of anxiety produces the stress response which is psychophysiological in nature and which has both short- and longer-term effects.

When anxiety becomes associated with a persistent and excessive fear of an object or situation which is not normally dangerous, a phobia can be said to exist (Butler 1989). The person's attempts to control the situation often involve avoiding it at all costs, but when that situation is encountered uncontrollable distress response occurs.[4] Even the thought of the stress-inducing event can produce the anxiety symptoms[5] and eventually the fear can dominate the person's life to such an extent that everyday functioning becomes almost impossible. The person feels they have lost control and that something terrible is going to happen to them if they do not escape from the fear-invoking situation. This can lead to acts of uncontrolled panic which often make the situation worse and reinforce the fear of disaster (Figure 36.1).

Cathy appeared to have developed a social phobia: her anxieties were persistently associated with the fear of humiliation and embarrassment in the presence of others. This disorder is serious, relatively common and affects men and women equally (Barondes 1993).

---

[4] Refer to pp.639–40 and to Table 22.2 to distinguish between eustress and distress.
[5] Refer to p.640, Table 22.1 for some psychophysiological indicators of the stress response.

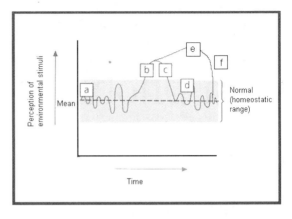

Figure 36.1 Phobic anxiety. (a) Subconscious perceptions of life events and daily hassles which equate to the healthy being. (b) Conscious perceptions of life events and daily hassles outside the parameters which equate to normal pattern behaviour. (c) Altering self-perceptions naturally by increasing self-awareness and behaviour modification. (d) Return of perceptions of life events and daily hassles to 'fit in' with normal behaviour patterns: sustained by subconscious appraisal. (e) Phobic behaviourism caused by the failure of reappraisal mechanisms. (f) Intervention to facilitate the reappraisal of phobic behaviourism with psychotherapeutic techniques such as opperant conditioning, etc.

Care

The treatment of a phobia depends on identifying the precise nature of the source of anxiety and relating it to specific goals which can be clearly measured. Often the person is asked to grade their anxiety-producing situations in order of severity. This produces a hierarchy of fears which can be used to assess the severity of the phobia, guide goal-setting and measure progress during treatment.

The most effective form of treatment appears to be through graded exposure to the anxiety-provoking situation under the guidance and support of a therapist. The early stages often involve teaching relaxation skills so that the patient is able to apply them to gradual, controlled exposure. Starting with the least fear-inducing situation the therapist can help the patient work toward developing control in circumstances which in the past have caused great anxiety. Success in graded stages builds confidence in the ability to cope (Figure 36.1). The exposure can be real or imagined but eventually the aim must be to see the patient approach the originally intolerable situation with the confidence that they will remain in control.

Phobic anxiety is an interplay of psychobiosocial events. The psychological perception of a threat produces the physical symptoms which the person then associates with social experiences. When the perception of threat becomes associated with something which is generally harmless the sufferer has usually learned this response and their treatment would be directed at a process of learning new, more positive responses to the situation which has caused the fear. Cathy could not control her fears of impending disaster because she had set in motion a number of physiological reactions which were beyond her conscious control. It would be futile to attempt to teach Cathy to stop her blood pressure increasing or her heart from pounding or her palms from sweating once the stress response was under way. Treatment was oriented toward allowing her to learn how to control her *psychological perception* of the event rather than her *physiological response* to it.

This type of treatment has been shown to be very successful and many people have been able to regain confidence and control in situations which had previously evoked panic and terror.

Further information

The most common phobia for people seeking professional help is agoraphobia: fear of open spaces. This accounts for 60 per cent of all phobic patients (and about 6 per 1000 of the general population). Social phobia such as that experienced by Cathy is the next most common phobia, accounting for 8 per cent. Zoophobia – animal phobia – accounts for a further 3 per cent.

The prevalence of specific fears and phobias change with age. The phobias associated with injections, doctors, darkness, snakes, heights, enclosed places and social situations show a dramatic increased incidence from childhood to late adolescence. The incidence of such phobias then shows a decline during adulthood, reaching their lowest incidence in old age. However, not all phobias follow this pattern. For example, the phobia associated with death, injury and illness demonstrates an increase from childhood to approximately 30 years of age. During mid-life there is a decline in the incidence of such phobias, only to rise again during one's fifties and sixties, with a sharp decline after that.

Main reference

Clancy, J., and McVicar, A. J. 1995 *Physiology and anatomy: a homeostatic approach*. London, Edward Arnold.

Other references

Barondes, S. 1993 *Molecules and mental illness*. New York, Scientific American Publications.

Butler, G. 1989 Phobic disorders. Cited in Hawton, K., Salkovskis, P., Kirk, J. and Clark, D. (eds). 1989 *Cognitive behaviour therapy for psychiatric problems – a practical guide*. Oxford, Oxford Medical Publications.

37

The case of a man with suicidal ideation

Sally Hardy

Learning objectives

1 To identify the different theoretical approaches to understanding suicidal behaviour

2 To discuss the implications for nursing a suicidal patient within a hospital setting

3 To explore the concept of suicide and aspects of mental health

Case presentation

Thomas Wright, aged 53, was admitted to a psychiatric hospital. He had been found in his garage, sitting in his car with a hosepipe attached to the exhaust. The pipe was stretched into the car and stuck with tape around his face. A neighbour found Mr Wright by accident: he had come to deliver some misdirected mail. The police and an ambulance were called.

Mr Wright had recently been made redundant from work, where he had been a sales manager. His wife and daughter were out shopping and were not due home until later that day.

Case background

Many studies have tried to establish the cause of and/or reason for suicide within society. The French sociologist Emile Durkheim first published his study of suicide in 1897 (translated into English in 1952). It has become a classic text. Durkheim's work linked suicide with a lack of social integration (Figure 37.1). Prior to this work, suicide was considered a result of an abnormality in the brain. Since 1897, many research studies have tried to understand suicide from many different approaches. These fall mainly into the following categories:

1 sociodemographic, e.g. age, gender, culture

2 behavioural, e.g. drug/alcohol abuse

3 developmental psychology and/or psychiatry, e.g. mood changes, personality

4 rational, e.g. social network deficiencies (Keinhorst *et al.* 1991)

Barraclough, Bunch and Sainsbury (1974) studied 100 cases of suicide. They concluded that 90 per cent were associated with mental illness, in particular alcoholism, or depressive states. (See Chapter 35's case study of a man with depression.)

Suicide rates have been found to be higher among males (particularly those under 30

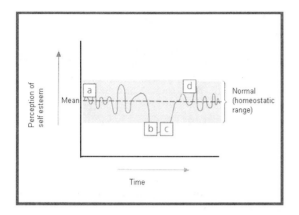

Figure 37.1 Suicide. (a) Self-esteem equates to that expected by societal norms and values. (b) Reduced self-esteem through a lack of social integration, which predisposes the individual to palliative coping mechanisms such as alcohol abuse and, in a severe case, to suicide. (c) Protecting the individual: initially interventions are focused on ensuring a secure environment. Later, pharmacological and counselling therapies act to restore self-esteem. (d) Self-esteem restored.

years of age). The major diagnostic categories are schizophrenia and affective disorders (see Chapter 33's case study on schizophrenia; Barraclough, Bunch and Sainsbury 1974; Copas and Ashley 1982; Kelleher and Daly 1990; Pritchard,1992; Charlton *et al.* 1993).

There is a wide coverage of suicide within the literature but it is argued by Taiminen (1992) that emphasis has been on macro studies rather than looking to the micro studies of smaller communities. Taiminen calls for further research and investigation into the mind's inner world.

Care

The standard nursing approach invariably entails assigning the patient to close surveillance, the removal of any harmful objects – even personal possessions – and restricting movement on the ward, all in an attempt to prevent further self-harm.

Although precautions vary from hospital to hospital, the aim is to maintain the patient's life within a safe and secure environment.

The patient's level of safety is immediately assessed. This includes consideration of factors known to represent an increased potential for suicide. Hradeck (1988) suggests three additional areas to be assessed:

1   that of the precipitating factors leading up to the crisis[1]

2   the patient's strengths and abilities to use coping mechanisms[2]

3   the nature and strength of the patient's support systems, such as family and friends

Apart from this custodial approach an interpersonal approach has been described as more therapeutic, where staff and patients concentrate on developing a working relationship (Bustead and Johnston 1983). The development of an interpersonal relationship is sustained and enhanced through the use of one-to-one communication sessions, where the patient and nurse meet together regularly to discuss issues of concern and begin to piece together a treatment programme. The nurse's aim is to help the patient come to terms with some very difficult and painful emotional and psychological problems.

The patient's family are encouraged to be involved with the treatment plan and to express any concerns of their own about the patient's condition.

Medication is often used, in particular antidepressants: in this case trifluperazine (stelazine) was prescribed. Antipsychotic drugs may be appropriate for some patients and are considered to act by interfering with or blocking the dopamine receptors[3] in the brain (Figure 37.1).

All referals are from the main reference at the end of this case study.
[1] Refer to pp.642–6 for a discussion of the subjectivity of stressors.
[2] Refer to pp.652–6 for a discussion of coping mechanisms and features of a failure to cope.
[3] Refer to p.367 for the action of inhibitory neurotransmitters such as dopamine.

Further information

Suicide and death remain strong taboos in our society. Suicide carries with it a variety of meanings and associations, most of which can be traced back to people's fear and anxiety. Freud (1950) identified and described two basic drives in man: the 'eros' (life instinct) and 'thanatos' (death instinct). These drives are restrained by social and cultural beliefs, which leads to frustration, aggression and feelings of guilt and fear. Anger is repressed and internalized in a maladaptive way, which can lead to self-destruction. Neurophysiologists have identified the limbic system[4] and in particular the hypothalamus[5] as being connected with violence and aggressive behaviour.

Some people remain firm in the belief that suicide is a symptom of all that is wrong in society, with the implication that the problem could be resolved through social enlightenment. Yet there is little evidence to support this (Alvarez 1971; Waern *et al.* 1996).

The real motives which impel a man to take his own life are elsewhere; they belong to the internal world, devious, contradictory and labyrinthine and mostly out of sight. (Alvarez 1971).

Main reference

Clancy, J, and McVicar, A. J. 1995 *Physiology and anatomy: a homeostatic approach*. London, Edward Arnold.

Other references

Alvarez, A. 1971 *The savage god. Study of suicide*. Harmondsworth, Penguin Books.

Barraclough, B., Bunch, J. and Sainsbury, P. 1974 A hundred cases of suicide: clinical aspects. *British Journal of Psychiatry* 125, 355–73.

Bustead, E. L. and Johnston, C. 1983 The development of suicide precautions for an inpatient psychiatric setting. *Journal of Psychosocial Nursing. Mental Health Services* 21, 15–19.

Charlton, J., Kelly, S., Dunnell, K., Evans, B. and Jenkins, R. 1993 Suicide deaths in England and Wales: trends in factors associated with suicide deaths. *Populations Trends*, Spring 71, OPCS 34–42.

Copas, J. B. and Ashley, R. 1982 Suicide in psychiatric inpatients. *British Journal of Psychiatry* 141, 503–11.

Durkheim, E. 1952 *Suicide. A study in sociology*, (ed) Simpson, G. London, Routledge.

Hradeck, E. A. 1988 Crisis intervention and suicide. *Journal of Psychosocial Nursing* 26, 24–7.

Keinhorst, C. W. M., DeWilde, E. J., Diekstra, R. K. W. and Wolsters, W. H. G. 1991 Construction of an index for predicting suicide attempts in depressed adolescents. *British Journal of Psychiatry* 159, 676–82.

Kelleher, M. J. and Daly, M. 1990 Suicide in Cork and Ireland. *British Journal of Psychiatry* 157, 533–8.

Pritchard, C. 1992 Is there a link between suicide in young men and unemployment? A comparison of the UK with other European community countries. *British Journal of Psychiatry* 160, 750–6.

Taiminen, T. J. 1992 Projective identification and suicide contagion. *Acta Psychiatrica Scandinavica* 85, 449–52.

Waern, M., Bekow, J., Runeson, B. and Skoog, I. 1996 High rates of antidepressant treatment in elderly people who commit suicide. *British Medical Journal* 313, 1118.

---

[4] Refer to p.372, and to Figure 13.21 for the location of the limbic system.
[5] Refer to p.373, and to Figure 13.23 for the location of the hypothalamus.

The case of a woman with post-puerperal psychosis

Sally Hardy

Learning objectives

1 Identify the precipitating factors leading to the onset of puerperal psychosis

2 Explore some of the theoretical explanations for postnatal psychological disorders

3 Discuss the implications of nursing a mother and baby in a hospital setting

---

Case presentation

Melissa Carter is a 33-year-old woman. She works as a local council administrator and has recently had a normal healthy baby. She was admitted to the mother and baby unit in a psychiatric hospital because her behaviour was becoming increasingly bizarre. Her GP and health visitor were concerned for the physical health and safety of both mother and baby.

Melissa had been talking about her new baby daughter as the new Messiah and had decorated her flat with religious artefacts. The health visitor had asked the GP to visit Melissa at home as she appeared not to be coping with the new baby.

Melissa's partner was also concerned as she was preventing him having anything to do with the baby and even stopped him entering the flat. Her parents lived abroad and had not been informed about her pregnancy as Melissa was convinced her parents would not approve.

On admission, Melissa was quiet and looked frightened. She held the baby very tightly in her arms and would spit at anyone who tried to interact with her or mention the baby by name. She remained in her room and would only allow the nurses near the baby when she was eating or in the bath.

Case background

Postnatal depression, or the 'baby blues', affects a majority of women and is considered a normal experience in the first few days after delivery. Most women are tearful, anxious, overwhelmed and miserable, following the traumatic yet exhilarating experience of childbirth. It is of concern to health professionals when these 'normal' reactions continue unabated and the mother slips into full-blown clinical depression (Figure 38.1). (See Chapter 35's case study on depression.)

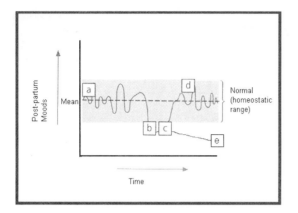

Figure 38.1 Puerperal psychosis. (a) Mood swings within acceptable limits after childbirth. (b) Mood swings in the first week or two after giving birth beginning to 'dip' excessively, indicative of swings associated with depressive states . (c) Intervention, mainly focused on fostering mother–baby bonding. Also attendance at support groups, and possibly the use of anti-psychotic drugs. (d) Re-established mood state according to expected pre-partum norms. (e) Clinical depression as the result of a failure of intervention to support the mother's mood in the first week or two after childbirth.

On the other hand, post-puerperal psychosis is considered a relatively rare and probably the most traumatic of any of the post-partum psychological disorders. Restlessness and disturbed behaviour normally appears a week or so after delivery. Kumar *et al.*'s (1995) study of affective psychosis identified that admission generally occurs within the first two weeks after delivery.

Historically, most psychotic episodes following childbirth were caused by toxaemia, while other emotional changes were put down to 'milk fever' and associated with hormonal changes. There is continued interest in the endocrine causes for postpartum disorders, although research continues to confirm a definite link (Davis, McIvor and Kumar 1995).

Care

The nurse's priority is to gain Melissa's trust and assure her that both she and her baby are being appropriately cared for. The need to observe Melissa caring for her child and to observe their developing relationship allows the nurse to assess Melissa's mental state and her ability to care for her baby (Figure 38.1).

On the mother and baby unit, Melissa was given the opportunity to join the other mothers to discuss their experiences in support groups, and to spend time in individual interaction sessions with nurses.

The multidisciplinary team meet regularly to discuss their patient's progress and often prescribe antipsychotic drugs (haloperidol or chlorpromazine) during the acute phase of the psychosis. As the acute stage is generally quickly abated, the medication is then changed to a mood-stabilizing drug, such as lithium or carbamezapine. Following this course of treatment some mothers are then discharged on antidepressants.

Melissa's partner was kept informed and involved with the treatment regime and encouraged to spend time with both her and their baby. When appropriate, he was able to take them both out for structured periods of leave from the ward.

Further information

Described as the most worrying form of perinatal psychiatry, puerperal psychosis is seen in about one in 600 pregnancies. It is typical in women who have not had any previous mental illness, and usually with their first baby. The symptoms begin with rapid mood swings, disturbed behaviour, restlessness and difficulty sleeping. This culminates in misperceptions, an inability to cope with the demands of a new baby, invariably auditory hallucinations (hearing voices) and other general features of psychosis (losing touch with reality).

Research has tended to concentrate on finding a precipitating factor for puerperal psychosis, looking to endocrine imbalance and following the theme of hormonal and biochemical imbalance (Riley 1979). But there is increased interest in a more eclectic approach and the link between individual precipitating factors, such as the lack of social support, social conditions, stressful life events and the transition of women to motherhood.

Ussher (1992) argues that women have been rendered unstable throughout history because of their menstrual cycle. This is now enlisted in the psychiatric textbooks as a cause of illness and an entity in itself (postnatal disorder, premenstrual syndrome) yet there is contradictory empirical evidence to support the menstrual cycle link with a fluctuating mental state.

Winnicott (1988) explored the early mother–child relationship and tried to encapsulate its complexity in the word 'hold'. The mother is expected to hold the baby, not just physically in her arms, but also psychologically and emotionally. When the process is broken (by the mother being unwell, for instance) it is the mother who needs extra care and holding, to encourage her to restore the role required of her by the ceaseless demands of a new baby. He concludes that those who are in the position of caring for a baby are as helpless in relation to the baby's helplessness as the baby can be said to be[1]

References

Davis, A., McIvor, R. J. and Kumar, R. C. 1995 Impact of childbirth on a series of schizophrenic mothers: a comment on the possible influence of oestrogen on schizophrenia. *Schizophrenia Research* 16(1), 25–31.

Kumar, R., Marks, M., Platz, C. and Yoshida, K. 1995 Clinical survey of a psychiatric mother and baby unit: characteristics of 100 consecutive admissions. *Journal of Affective Disorders* 33(1), 11–22.

Riley, D. 1979 Puerperal psychiatric illness: psychogenic or biochemical? *Proceedings of the Fifth World Congress of the International College of Psychosomatic Medicine.*

Ussher, J.M. 1992 *Women's madness: misogyny or mental illness?* London, Harvester Wheatsheaf.

Winnicott, D.W. 1988 *Babies and their mothers.* London, Free Association Books.

39 The case of a woman with Huntington's disease

Andrew McVicar and John Clancy

> **Learning objectives**
>
> 1 To revise the role of the brain nuclei involved in controlling movement
>
> 2 To understand the neurological changes caused by the Huntington's gene
>
> 3 To be aware of the care provided for an individual with a progressive neurological disorder of this type

Case presentation

Jane is 38-years-old. On a recent visit to her GP for a routine check-up, her doctor was concerned to see that Jane frequently, and apparently unintentionally, smacked her lips and grimaced slightly. She was also irritable and fidgety. The doctor was aware of Jane's family history: her father had died a few years earlier of Huntington's disease. He referred her to a neurologist for a further examination. In view of her age, family history and symptoms, the diagnosis was straightforward: Jane was exhibiting the early signs of Huntington's disease.

Case background

Huntington's disease is caused by an autosomal dominant gene.[1] How the gene operates is poorly understood but neurological changes are observed that typically include a disturbance of the functioning of the basal ganglia of the forebrain and the substatia nigra of the midbrain.[2] These brain nuclei are involved in the control of posture and movement. They are particularly involved in the control of the muscles of the face and of the manipulative ability of limbs. The gene responsible for Huntington's disease causes:

- a progressive loss of the inhibitory neurons which utilize the neurotransmitter gammaaminobutyric acid (GABA).[3] This disturbance of neurochemical homeostasis causes the symptoms exhibited by Jane
- a progressive decline, and eventual atrophy, of the brain nuclei leading to a

All referals are from the main reference at the end of this case study.
[1] Refer to p.29 and p.598 for a discussion of genes.

[2] Refer to p.371 and Figure 13.20 for the location of basal ganglia and to p.374, Figure 13.24 for location of substatia nigra.
[3] Refer to p.367, Table 13.3 for notes on GABA.

worsening of symptoms. Involuntary jerky movements of the head, arms and legs occur (the alternative name for the disease is Huntington's chorea and this relates to the dance-like or choreiform movements). Postural abnormalities develop with the shoulders pushed back and the lower trunk pushed forward

- as the disease develops, there is also a deterioration of the neural pathways to the frontal cortex[4] leading to a decline in cognitive functions and eventual dementia

Care

Huntington's disease is irreversible and its progression cannot be prevented. In the early stages mild tranquillizers may reduce the muscle contractions but these actions become progressively weaker. Therapy is directed at delaying admission to residential care for as long as possible. Care aims to:

- help the individual to communicate effectively
- prevent the patient from becoming isolated. The education of friends and relatives is important in this respect. Nurses must also contain their own reactions to sudden abnormal movements
- ensure that food intake is appropriate. In relation to dietary needs, it should be remembered that muscle contraction is an efficient means of generating energy.[5] Dietary energy content must be increased accordingly (see Figure 39.1). As feeding can become difficult and messy as the condition progresses, it is also important to ensure that food is provided in manageable pieces. A drinking straw may also help
- maintain personal hygiene as self-care diminishes

[4] Refer to p.370 for a description of what constitutes the frontal cortex.
[5] Refer to pp.58–60 for a discussion of energy production by cellular respiration.

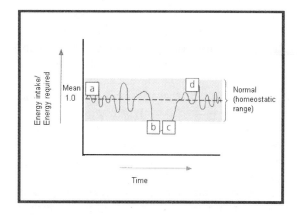

Figure 39.1 Huntington's disease. (a) Energy balance (energy requirement = dietary energy intake). Small fluctuations average out with time. (b) Disturbance in energy demands as energy requirement exceeds the energy available from the diet as a result of increasing (involuntary) muscle activity. Emaciation will result if dietary energy intake is not increased to restore the balance, as in (c). (d) Good control of energy balance, with complete compensation for increased energy demands.

- maintain personal safety
- reduce the suicide risk. Daycentre care, voluntary visitor schemes and means of entertainment are important ways of reducing depression
- support the patient's family
- provide genetic counselling for the patient's children (see later)
- provide psychological care to facilitate the acceptance of the changing self, frustration and the terminal nature of the disease

The rationale for most aspects of the care provided in Huntington's disease is self-evident, bearing in mind the increasing supervision Jane will need as the disorder progresses.

Further information

Huntington's disease is autosomally inherited, so it affects both sexes equally.

Although caused by a dominant gene mutation, it does not usually have its onset until the individual is 30–45 years of age. Spontaneous intrauterine mutation of the gene is a very rare occurrence, so the gene is usually inherited; the dominant nature of the gene means that a parent probably also had the condition. The late onset of the disease is not understood, but it has implications for any children the affected Huntington's individual may have.

With one affected (heterozygous) parent there is a 50 per cent risk that their offspring will inherit the gene.[6] Jane's children, therefore, will have to consider the possibility that they too might eventually develop the disease, and may one day also pass on the gene to their children. Genetic counselling is important.

The Huntington gene has now been identified (on chromosome 4) and predictive gene-testing may become widely available. This still presents a dilemma for children of an affected individual as 'knowing' can be worse than 'not knowing'. It does raise the possibility, however, of embryo-screening in *in vitro* methodologies.[7]

Main reference

Clancy, J. and McVicar, A. J. 1995 *Physiology and anatomy: a homeostatic approach*. London, Edward Arnold.

---

[6] Refer to pp.603–4 for a description of autosomal dominant inheritance.

[7] Refer to pp.609–10 for notes on embryo-screening, genetic-engineering and gene therapies, and genetic-fingerprinting.

PART FOUR

Learning disability case studies

The case of an adolescent girl with epilepsy

40

Gabriel Ip

Learning objectives

1 To understand the basis of epilepsy as a symptom and not a disease

2 To recognize the rationale for therapeutic intervention in epilepsy

3 To appreciate the holistic and intradisciplinary characteristic of care

Case presentation

Jenny is now 18 years old. She was driven to a party one evening two years ago with her boyfriend, George. George had been drinking. There followed a serious accident and George was killed instantly. Jenny was found unconscious and was rushed to the local hospital. She had suffered extensive facial and head injuries. At one point, she was given a very poor prognosis because of her cerebral damage and subarachnoid haemorrhage. Miraculously she pulled through, but she was left with paraplegia and needs to be confined to her wheelchair.

Jenny has regained most of her other motor skills though there is a slight slurring in her speech. She has been having 'hallucinations' of flashing lights and experiencing a bitter taste. This is followed by blackouts quite regularly throughout the day.

Last week Jenny started to exhibit involuntary convulsions of the whole body instead of the blackouts. There were first stiffening and followed by powerful rhythmic jerking of her limbs lasting one to two minutes. There were noticeable changes in her consciousness, behaviour and perception (reportedly) and her parents who were at home with her at the time had become very anxious indeed. After discussions with her GP, Jenny was referred to a neurology specialist.

Jenny has had two sessions of electroencephalograms (EEG) investigation performed in a week. Her EEGs demonstrated bilateral high voltage delta[1] activity over the affected (injured) area (Figure 40.1).

Jenny is subsequently diagnosed as having generalized tonic-clonic epileptic seizures caused by organic brain damage.

Case background

Post-traumatic generalized epileptic seizures is not uncommon among young people

All referals are from the main reference at the end of this case study.
[1]Refer to p.380 for a definition of delta waves.

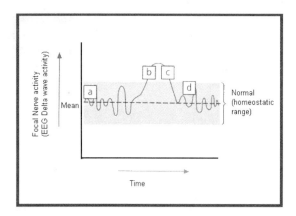

Figure 40.1 Epilepsy. (a) Contained electroencephalogram (EEG) δ-wave activity in the cerebral cortex associated with health. (b) Excessive EEG δ-wave activity in the cerebral cortex as experienced with epilepsy. (c) Intrinsic control via 'dampening down' EEG δ-wave activity in the cerebral cortex as occurs with time following a 'fit or seizure'. (d) Extrinsic control of EEG δ-wave activity in the cerebral cortex using anti-seizure drugs and psychosocial care.

especially after some forms of organic brain damage. The result of the damage leads to a decreased pathway activity caused by the loss of neurones within the pathway.[2] This in turn leads to excessive inhibitory synaptic activities relative to the inhibitory synaptic activities.[3] The onset is usually slow but progressive: that is, should the condition fail to be brought under control, the length of seizures would widen and the risks of sustaining physical injuries expand.

In Jenny's case, her epileptic seizures are caused by an acquired organic cerebral lesion through the serious accident she had. Although she has made a dramatic recovery from her injuries, the cerebral lesion remains. This, in turn, causes a paroxysmal discharge of abnormal electrical rhythms at the lesion sites.

[2] Refer to p.393 for notes on the loss of neurons which in turn leads to excessive inhibitory synaptic activities relative to the inhibitory synaptic activities.
[3] Refer to pp.395–6 for notes on the imbalance between inhibitory and excitatory neurotransmitters.

These abnormal electrical rhythms cause Jenny to have:

- involuntary muscular convulsions
- altered perceptions
- altered level of consciousness

Muscular convulsions are the results of abnormal electrical discharges (motor). Altered perceptions are the results of abnormal electrical discharges (sensory) and the altered level of consciousness is the result of interference in the brain in certain areas that may or may not lead to unconsciousness (Bannister 1978).

Care

Since the causal lesion to the cerebral cortex cannot be reversed, the approach to care is a symptomatic one. That is, the management and control of further seizures (this includes preventive and inhibitive interventions) would then be the prime concern.

In Jenny's case, her epileptic seizures are residual to her cerebral lesion. Therefore the emphasis of intervention must be firmly based on symptomatic treatment. However, Jenny's needs are multi-faceted. The management of her seizures lies in the crossroads between that of general medical practice, emergency medicine, community and hospital-based nursing and medical care.

For an holistic approach to the management of epilepsy (Figure 40.1), one should advocate the following:

- an accurate diagnosis
- psychosocial care
- medical control
- the routine associated with drug treatment
- emergency medical care
- emergency first aid
- occupational therapy
- education
- the surgical removal of epileptogenic tissues
- an oligoantigenic diet

An accurate diagnosis must make use of the combined efforts of observations from those close to Jenny (time, frequency, duration and description of seizures, etc.), clinical examination and EEG. The electroencephalogram is only an aid to diagnosis and is valuable only when its results are considered jointly with the history and clinical examination of the patient. An area of delta waves indicates an area of organic brain damage (Bannister 1978).

Psychosocial care involves reassuring the patient as well as her parents and other informal carers. Fear, uncertainty and even superstitious folklore must be allayed by close counselling with Jenny and her carers. Individual assessment must be made about her choice of activities. The risks must be explained to her and her carers so that an informed choice can be made. It must be understood that it is impossible to guard against all risks even by depriving her of a 'normal' repertoire of leisure and social activities.

Medical control involves administering the prescribed medication. Certain drugs have been found to be valuable in negating the severity and frequency of epileptic attacks. Epilim (sodium valproate), epanutin (phenytoin sodium) and phenobarbitone are some common choice drugs in the control of generalized tonic-clonic seizures. The object of drug treatment is to secure an abolition of attacks for a sufficient length of time to enable the patient to lose the epileptic habit. When the seizures occur fairly frequently at the same hour of the day or time of the month, the dose could be timed correspondingly to produce its maximal effect. The role of the nurse or carer must be to ensure that the right drug is taken at the right time at the right dose through the right route to the right person.

The routine associated with drug treatment involves first the practice of monitoring serum drug toxicity. Since no drug is completely free of side-effects and its accumulative toxicity, it is imperative to make sure the patient is never overdosed by medication through albeit appropriate treatment. Blood samples are collected to evaluate therapeutic doses and toxicity.

Second, another aspect of care is the detection and treatment of drug-induced Parkinsonism since anti-epileptic drugs are mainly neuroleptics which can produce extrapyramidal symptoms.[4] Third, the maintenance dose needs to be stabilized and not suddenly completely withdrawn, as it may precipitate status epilepticus. This is not unheard of, especially in people taking barbiturates.

Emergency medical care is involved when the changing of personal and environmental conditions forces the patient to go into a series of seizures without regaining consciousness in the various stages of seizure (status epilepticus). This is life-threatening and hence considered to be a medical emergency. The nursing activities in this case would include a detailed careplan to help to prevent exhaustion, physical injuries and other complications such as cardiac arrest, hyperpyrexia, asphyxiation, aspiration pneumonia and death. Immediate admittance to a hospital may secure the prompt administration of IV anti-epileptics (e.g. valium or paraldehyde); bed rest; skin care; proper positioning; cooling by fanning and/or tepid sponging; cot sides to prevent rolling over; close observations and professional reassurance to parents; and detailed record-keeping. With proper medical and nursing care status epilepticus could be controlled.

Emergency first aid involves a day-to-day encounter when the patient suddenly falls over and develops a seizure. Checking the environment and rendering it safe is the first priority before approach. Remove the danger away from the patient or vice versa. The essentials of first aid in this situation is not what the carer does but rather what they do not do. In brief, the least done the better. Ensuring the patient has a clear airway is paramount (never introduce a foreign object into the patient's mouth to help breathing). Just loosen any restrictive clothing. Many injuries are the result of kind-hearted acts of restraining the person in seizure

---

[4] Refer to p.498 for an overview of the extrapyramidal pathways.

and the introduction of spoons, etc. into the patient's mouth.

Occupational therapy involves the day-to-day training and retraining of self-help skills and work skills. Professional advice on the selection of occupation is a pragmatic approach to rehabilitate the patient: there are some statutory barriers to some occupations, e.g. airline pilot, deep-sea diver and train-driver to name but a few. Other leisure and social activities have to be assessed on the patient's own circumstances.

Education involves not only the patient but also their carers. Irrational fears and myths must be dispelled at all costs. It is desirable that an epileptic patient should as far as possible lead a normal life. Once the seizures are brought under control, there should be no prevention of the patient attending school or following a vocation. Counselling and information-giving to the patient and their carers is all-important.

The surgical removal of epileptogenic tissues involves first the treatment of focal cerebral pathology when the accurate diagnosis has established a distinct link of the focus with seizures; or, second to influence the cerebral function to prevent or restrict the propagation of abnormal electrical discharge (Schwartz 1994). There is also increasing scope to apply stereotactic to destroy redundant but offending cerebral tissues. Resective operations are gaining popularity to correct temporal lobe epilepsy in children (Weiser 1986).

Oligoantigenic diet involves the detailed assessment of dietary intake by qualified dieticians and the prescription of low-allergenic meals. Wurtman (1983) suggests that the neurotransmitters involved might be derived from the gut and that the intracerebral level of amino acid-derived neurotransmitters can be affected by dietary precursors. Opiate-like peptides have been discovered in milk and wheat and implicated in the pathophysiology of epileptic seizures (Frenk, Engel and Aukerman 1979).

Further information

Ninety-five per cent of epileptic disorders are idiopathic with no observable cause. Most epileptic convulsions occur in childhood where the outlook is good. Ninety-five per cent of attacks of febrile convulsions cease by the age of five years and are not the precursors of epilepsy in adulthood (Ross and Woody 1994).

In adults with epilepsy, in 50 per cent the attacks cease or can be controlled, one-third will suffer one or two attacks in one year and only 15 per cent suffer more than two attacks in a year. Only 5 per cent of epilepsy sufferers are severely disabled.

Main reference

Clancy, J, and McVicar, A. J.. 1995 *Physiology and anatomy: a homeostatic approach*. London, Edward Arnold.

Other references

Bannister, R. 1978 *Brain's clinical neurology*, 5th edn. Oxford, Oxford Medical Illustrations.

Frenk, H., Engel, J., Jr and Aukerman, R. 1979 Endogenous opioids may mediate postictal behavioural depression in amygaloid kindled rats. *Brain Research* 167, 435–40.

Ross, E. and Woody, R. (eds) 1994 *Baillière's Clinical Paediatrics – International Practice and Research*, 2 (3). London, Baillière-Tindall.

Schwartz, R. 1994 Non-pharmacological approaches to children's epilepsy. Cited in *Baillière's Clinical Paediatrics*, London, Baillière-Tindall.

Weiser, H. 1986 Selective amygladalo-hippocampectomy: indications, investigative techniques and results. Cited in *Baillière's Clinical Paediatrics*, London, Baillière-Tindall.

Wurtman, R. 1983 Behavioural effects of nutrients. *Lancet* I, 1145–7.

The case of a man with Down's syndrome

41

Derek Shirtliffe

Learning objectives

1 To understand the cause of Down's syndrome

2 To appreciate the needs of people with Down's syndrome

Case presentation

WECHSLER ADULT INTELLIGENCE SCALE (WAIS)

Tony is 50 years old. He lives at home with his parents who are both elderly. Shortly after his birth Tony was discovered to have Down's syndrome. As a baby he exhibited hypotonia. His parents remember him as a 'floppy' baby who was quiet and passive throughout most of his early childhood. His IQ when measured at 12 years of age, using WAIS, was found to be 60: lower than normal. He received little formal education during his childhood and has attended a daycentre since his late adolescence.

Tony has poor eyesight and commonly experiences conjunctivitis and blepharitis. His visual disability is exacerbated by his frequent refusal to wear spectacles which have been prescribed for his visual impairment. Tony also has hearing difficulties. The combination of sight and hearing difficulties often have an adverse impact on Tony's attempts to communicate with other people and this is particularly acute when he tries to talk to strangers.

Throughout his childhood and adult life, Tony has been mildly obese. Recently he has become withdrawn and is described as apathetic by his parents and people who know him well. While previously he could undertake self-help skills such as bathing independently he now requires a considerable amount of help to perform these skills successfully.

Tony also seems to be confused by situations which he previously understood well, such as hobbies which he used to enjoy. His parents are particularly frightened by a recent incident during the night where Tony seemed to be confused by the darkened staircase leading to the ground floor when his parents were asleep. Tony fell down several of the steps but he suffered no serious ill-effects.

Case background

Down's syndrome is the most common chromosomal disorder causing learning disabilities. The identification of the condition is ascribed to John Langdon Down in 1866. Although a number of common anatomical and functional irregularities are found among people with Down's syndrome (Figure 41.1) they may vary greatly in their manifestation in

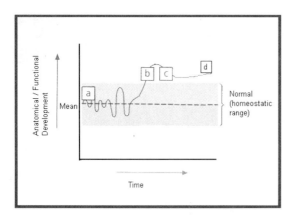

Figure 41.1 Down's syndrome. (a) Anatomical/ functional development of embryological tissues appropriate to the norms for embryological age. (b) Accumulation of anatomical and functional abnormalities in the embryo as a result of genetic or environmental influences. In Down's syndrome, such changes arise from a surplus of genes provided by an additional chromosome. (c) Limited correction of developmental abnormality after birth, for example, the surgical realignment of the blood vessels of the heart. (d) Persistent functional/anatomical abnormalities which cannot be corrected.

each individual sufferer. Some of the more common features are:

- brachycephalia with reduced cranial capacity. This largely accounts for the learning difficulty observed in Down's syndrome
- there may be an epicanthic fold on the inner aspect of the upper eyelid. Strabismus, nystagmus and cataracts are common. Brushfield's spots are flecked through the iris which may be poorly developed. There is often a reduced production of the enzyme lysozyme, which has antiseptic qualities, in tear secretion.[1] These changes account for Tony's visual problems
- the bridge of the nose is often small with a high narrow palate. The tongue is large with horizontal fissures and this leads to tongue protrusion with subsequent

All referals are from the main reference at the end of this case study.
[1] Refer to p.266 and p.269 for a detailed account of lysozymal action.

mouth-breathing. Such features contribute to the communication difficulties observed
- heart problems affect approximately half of all babies with Down's syndrome and historically this has led to many of their early deaths
- thyroid deficiency is much more common in people with Down's syndrome and it is responsible for a slowed metabolic rate.[2]

Tony's more recent lapses are indicative of the onset of Alzheimer's disease. Links with Alzheimer's disease and Down's syndrome are generally accepted as substantial (both involve chromosome 21). But the full extent of the association between the two conditions is not fully understood. It is generally accepted that Alzheimer's disease occurs more commonly in people with Down's syndrome than in the general populations and that onset is at an earlier age than in non-disabled populations. However, while there are significant pathological changes in the brains of many people with Down's syndrome consistent with Alzheimer's disease, this may not always be accompanied by the dementing behaviour of a pattern expected of this condition (Oliver and Holland 1986).

Care

The confusion which Tony experiences has caused his parents great concern and the case has recently been referred by their GP to a community learning-disability nurse. The community nurse's interventions with Tony are based on two main approaches. The first aims to enable Tony to cope more completely with his psychological deficits, such as failing memory and confusion. The second approach aims to change the environment in which Tony functions so that it better fits his diminishing cognitive capacities.

Alongside these objectives, the nurse recognizes the need to work with Tony's parents to

[2] Refer to p.441 for an account of thyroid hormones.

minimize their anxiety and to coopt them as partners in working to meet their son's needs. The nursing intervention seeks to recognize the impact on family routine, family leisure, family interaction and the general physical and mental wellbeing of Tony and his parents.

After a careful initial assessment, the nurse will work with Tony and his family on a number of priorities. First, the community nurse begins to work with Tony's parents to resolve their anxieties and fears associated with their son's changing behaviour. Some of their concerns are immediate, associated with the direct impact of the change on a daily basis, such as Tony's safety, given his confusion. With reference to longer-term issues their concern is primarily centred on Tony's welfare when they become too frail to care for him or after their deaths. The community nurse has begun to counsel both parents about these fears and some strategies such as parent-support groups and respite care have been suggested.

The community nurse has also begun strategies with Tony which aim to maintain his capacity to retain information via long-term memory. This has taken the form of using imagery in association with certain words, such as the concept of postman in association with Tony's favourite hobby of stamp-collecting. Another strategy adopted by the nurse is the use of reality orientation in conjunction with Tony's parents. Thus at every possible opportunity both Tony's parents reinforce with him notions such as the time of day, Tony's and their own names, and subjects such as forthcoming holidays. The community nurse will also give careful consideration to the apathy which Tony appears to display. For example, it might be agreed that Tony's father will encourage him to participate in activities they can enjoy together, such as gardening. With regard to restructuring the environment, Tony's bedroom within the family home will be moved to the ground floor of the house to avoid the risk associated with the stairs.

Further information

Down's syndrome arises because there is an excess of genes within the individual's cells. The syndrome is often alternatively called trisomy 21 because, in the majority of sufferers, the condition is caused by the non-disjunction of chromosome pair 21 during meiosis in the mother's gonads, leading to the presence of an extra chromosome in her gametes.[3] The incidence of this event increases as maternal age increases, presumably because of the ageing of the primary oocytes in the ovary (oocyte production begins before birth, with oocytes 'frozen' early in meiosis until development is retriggered during adulthood). Thus the incidence of Down's syndrome rises from 1/1600 when maternal age is 20–24 years to 1/46 when mothers are over the age of 45. A small number of cases are caused by translocation of all or part of an extra chromosome to another chromosome pair during meiosis.[4]

The overall incidence of Down's syndrome, incorporating influences such as increased maternal age, is often quoted as 1/600 live births but this is an approximation based on recorded incidence across several decades (Wishart 1988). As detection of Down's syndrome has been possible for several decades using methods such as amniocentesis, we might expect the incidence of the condition to have declined. However, contrary to this expectation, the impact of screening on the incidence of Down's syndrome has been slight (Steele 1993).

The reasons for this include:

- screening has commonly been limited to women over 35 years of age or women at special risk, such as those with a relative with Down's syndrome
- low rates of take-up of screening
- some women who are eligible for screening refuse on moral or religious grounds

---

[3] Refer to pp.606–7 for an explanation of chromosomal non-disjunction.
[4] Refer to p.607 for a description of translocation (chromosomal fragmentation).

Main reference

Clancy, J, and McVicar, A. J. 1995 *Physiology and anatomy: a homeostatic approach*. London, Edward Arnold.

Other references

Oliver, C. and Holland, A. J. 1986 Down's syndrome and Alzheimer's disease: a review. *Psychological Medicine* 16, 307–22.

Steele, J. 1993 Prenatal diagnosis and Down's syndrome: part 2. Possible effects. *Mental Handicap Research* 16 (1), 56–69.

Wishart, J. 1988 Early learning in infants and young children with Down's syndrome. In *The psychology of Down's syndrome* (ed) L. Nadel, 7–50, Boston, MA, MIT Press.

The case of a young man with sensory deprivation: the use of the multi-sensory environment

42

Derek Shirtliffe

Learning objectives

1 To review the way in which the body senses the external environment

2 To appreciate the application of the multisensory environment with people who have learning disabilities

3 To identify the special leisure needs of people with learning disabilities

---

Case presentation

Peter is 30 years old and lives at home with his parents. An assessment of his adaptive behaviour reveals that he has a severe learning disability and has no verbal language. He receives daycare through a health-care trust and nurses working with him have observed that he has an affinity to the multisensory environment they have developed. Since Peter has no spoken language, the nurses involved in his care have based their approach on careful observation of his behaviour, helped by the trust and rapport established over several months of relationship-building.

Case background

A 'sense' is a faculty by which our body receives information about the changes in its internal and external environments. The body detects changes by way of receptors. Some receptors are sensitive to a specific stimulus, such as touch, pressure and temperature, while others have the capacity to detect different types of stimulus. These more complex receptors are usually organized into sense organs, for example, the eye, ear and nose.[1]

All referals are from the main reference at the end of this case study.
[1] Refer to pp.399–407 for an overview of receptor types in the body.

Nurses working with clients who have learning disabilities have recognized the importance of enabling the development of learning via the construction of opportunities for clients to use their senses to interpret the environment which surrounds them. With this in mind, the services for people with learning disabilities have increasingly paid more attention to the leisure needs of clients. While the models of normalization dictate that these needs should be met via services in the wider public domain, there has been an acknowledgement of the special needs of people with learning disabilities. This has lead to the development of multisensory environments.

The origins of the multisensory environment can be traced back to the use of 'snoezelen' in The Netherlands. This word is a combination of two Dutch words: the equivalents in English are to 'sniff' and to 'doze'. While the original concept is associated with relaxation, the multisensory environment can include the application of this notion but it also includes more interactive activity through which learning can take place.

Care

The multisensory environment which Peter uses contains a number of different areas which are constructed so as to primarily stimulate a single sense while affecting others at a lower level (see Figure 42.1). The rooms are as follows:

- *Room one* has dimmed lighting and soft furnishings, including a water bed, which encourage relaxation. Soft music is played which accentuates the calming effect of the room. The desired effect of this room is that it provides an environment where environmental stressors are relatively few. Thus the individual is encouraged to assume a physiological, psychological and sociological condition whereby the learning opportunities of this and the other rooms are more likely to be accessed

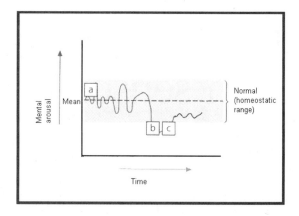

Figure 42.1 Sensory deprivation. (a) Normal level of arousal in people without learning difficulties. (b) Reduced level of arousal promoted by learning difficulty. (c) Exposure of the individual with learning difficulty to a sensory stimulus e.g. visual, olfactory, auditory, gustatory, to increase arousal to as near normal as is possible. Stimuli may be provided with a multisensory environment that provides the full range of stimulation.

- *Room two* contains a floor surface with areas which activate sounds when they are stepped upon. Other aspects of this room include wall-mounted panels which emit light when activated by sounds such as the human voice. This room gives Peter the opportunity to develop skills associated with the discrimination of tone, pitch and frequency.[2]
- *Room three* has soft furnishings and windows that have coloured glass which makes the exterior appear unusual and illuminate the interior with a warm glow. In this room visual impact is provided by the use of tall transparent tubes filled with water through which coloured bubbles percolate. The walls have a number of collages and a projector can be used to illuminate the room with interesting patterns and designs. A fibre-optic tail is also in the room which undergoes a succession of changes of colouration. Other effects are possible via the use of mirrors and

[2] Refer to pp.405–6 for a discussion of the 'special senses'.

lighting set within the walls. Within this environment Peter is likely to experience a variety of stimuli. The environment encourages the use of discriminatory skills such as the appreciation of light and shade. This is possible because the receptor cells devoted to the detection of visible colour (the cones) and those sensitive to low light intensities (the rods) will be influenced by the differing conditions within the room. The organization of the stimuli involves both the eye and the brain.[2]

- *Room four* is filled with materials which provide interesting tactile experiences. These include the use of materials with soft, smooth and rough properties. Some of the materials are ordinary materials such as pasta shells and spirals. These have the advantage of being safe if Peter explores their properties by putting them in his mouth. The materials chosen for this activity could be selected so as to provide the taster with an opportunity to experience the four basic tastes of sweet, sour, bitter and salt. These are discernible by means of receptors situated mainly on the surface of the tongue, but also distributed on the epiglottis, pharynx and palate.[3] The room also includes a ball pool which gives the user the possibility of enjoying safe movement through an unusual medium. Here Peter is able to gain experience of mass, density, texture, temperature and dimensions of objects. This is achieved by the body by means of receptors throughout the body which convert the energy within the perceived stimulus into electrical energy[4] which can be predicted by the brain.

- *Room five* contains a number of means for the user to experience the different sensations associated with smell. A number of tubes attached to the walls contain harmless substances which convey a range of smells. By using this equipment Peter can refine his discriminatory skills associated with the sensation of sweet or acrid smells and also the difference between subtle or pungent smells. His body can distinguish these differences by means of olfactory receptors within the mucous membrane of the nasal epithelium which send information to the olfactory bulb and subsequently to the limbic system.[5]

Further information

The multisensory environment offers Peter a number of advantages as a learning platform. It gives him a range of opportunities to develop discriminatory skills and the heightened arousal of his senses should ensure an increase in the likelihood of learning taking place. A number of other issues associated with the development of his independence are implicit in this style of care. First, he can exercise control over his own activities, thus developing decision-making and choice-making skills. This is an important issue for people with learning disabilities. They may be particularly vulnerable to exposure to institutional practices because of their perceived incapacity to articulate their needs and aspirations.

Second, the multisensory environment can be used as a vehicle for discovery at the individual pace of the client. Third, the level of sensory arousal can be controlled by the client so that the level of 'heating up' and 'chilling out' can be determined by the user with regard to duration, frequency and pitch.

Another consideration is that Peter has a poorly developed sense of danger and all the exploratory activity within the multisensory environment can remain loosely supervised, thus ensuring maximum independence, but

[2] Refer to pp.405–6 for a discussion of the 'special senses'.
[3] Refer to pp.420–1 for notes on the interpretation of gustatory signals.
[4] Refer to pp.363–5 for a discussion of electrochemical conduction in neurons.

[5] Refer to p.422 for notes on the interpretation of olfactory information.

can also be regularly checked for safety. Nurses using multisensory environments with clients intend to achieve either the alleviation of stress or heighten the arousal of the senses.

Main reference

Clancy, J, and McVicar, A. J. 1995 *Physiology and anatomy: a homeostatic approach*. London, Edward Arnold.

The case of a girl born with cytomegalovirus inclusive body disease

43

Christine Nightingale

Learning objectives

1 To understand the cause of cytomegalovirus inclusive body disease

2 To examine the impact on both the mother and child's homeostasis

3 To understand the rationale for care in relation to disordered physiology

Case presentation

Maria was born after a relatively uneventful first pregnancy. Generally her mother Cassandra was well, although she recalls feeling 'a little under the weather' at the end of the first trimester of pregnancy. She did not find this remarkable as she frequently compared notes with other pregnant women who also complained of fatigue.

Maria was diagnosed as having jaundice in the first 24r hours and was later observed to have infantile spasms. Her development was frequently monitored during her preschool years, as her head measurements showed that she was microcephalic and her infantile spasms had developed into tonic-clonic epileptic seizures.

Maria received full-time education until she was 19 in a school for children with special needs, as she was assessed as having severe learning disabilities. Because of her hyperactivity, lack of concentration and poor eyesight,

Maria places herself at great risk by wandering out of buildings and stepping out into the road without any recognition of danger.

Maria is now 22 years old. Caring for her put Cassandra's marriage under great strain and eventually Maria's parents separated with animosity. Now alone, Cassandra is exhausted from caring for all of Maria's needs and from constantly watching her to ensure her wanderings do not place her in danger.

Case background

The cytomegalovirus (a 'salivary gland virus') is one of the herpes family of viruses. The symptoms of cytomegalovirus may be slight and often go unnoticed. Typical symptoms may resemble glandular fever. They include malaise, fever, swollen lymph glands and abnormal liver function (Glanze 1986). Individuals with reduced or compromised immunity, such as babies, the elderly, those

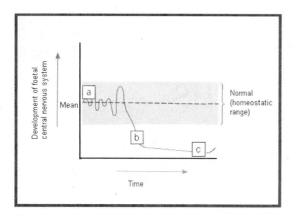

Figure 43.1 Cytomegalovirus. (a) Development of the fetal central nervous system at rates within the normal range appropriate to fetal age. (b) Diminished fetal development arising from the presence of a teratogen, such as cytomegalovirus, leading to learning difficulty. (c) Permanent microcephaly. Although some functional improvement can be promoted using, for example, relaxation therapies, the effect of intervention is very limited.

with HIV infection and AIDS and those being treated with immunosuppressive drugs are at high risk of suffering life-threatening illnesses such as pneumonia. The developing fetus is also at risk from maternal infection via the placenta,[1] particularly when the mother is suffering from a primary infection. As with maternal rubella, the fetus is at greatest risk of damage in the first trimester. However, unlike maternal rubella, infection of the fetus may not occur until much later in the pregnancy.

The effects on the infant, who is often preterm and with low birth weight, may be severe (see Figure 43.1). Typical physical features include microcephaly (small head, cranial circumference of 43 cm or less in an adult (Clarke and McCree 1985), severe damage to the nervous system resulting in epilepsy often presenting first as infantile spasms (West's syndrome) (Craft, Bicknell and Hollins 1986), cerebral palsy, deafness and hyperactivity (Clarke and McCree

All referals are from the main reference at the end of this case study.
[1] Refer to pp.588–9 for a definition and identification of common teratogens.

1985), and enlargement of the spleen and liver. A wide range of learning disabilities may be present from mild to severe. For the first few months the liver and spleen may be enlarged, resulting characteristically in jaundice. Typically the eyes are abnormally small (microphthalmos) and the individual often suffers poor vision (Glanze 1986).

Care

There is currently no known treatment for the cytomegalovirus condition. Health and social carers and educationalists can only respond to service-users like Maria in a sympathetic and humane way. Care plans must involve extensive programmes which aim to increase the concentration span and to make constructive and meaningful use of the person's energy and hyperactivity by recognized, valued and enjoyable exercise and activity. Conversely, service-users such as Maria need to be encouraged to relax; therapeutic areas such as multisensory environments ('snoezelen': see Chapter 42's case study of the young man with sensory deprivation: the use of the multisensory environment for a discussion of 'snoezelen') may be beneficial in encouraging interest, concentration and relaxation.

As with any epileptic condition the status of the epilepsy requires regular monitoring. Anti-seizure drugs such as sodium valproate will be administered and a vigilant awareness of the side-effects of these drugs must be maintained. The side-effects of sodium valproate include abdominal pain, hair loss, weight gain and skin rashes. Particularly relevant in Maria's case is that prolonged use of sodium valproate may cause liver damage, so regular liver-function tests are important.

Equally important is caring for the carer who may be looking after a highly active individual. In Maria's case, her mother Cassandra is now caring for her virtually alone. The care of difficult-to-manage individuals may result in social isolation, as others find the behaviour of service-users such as

Maria difficult to understand and find their constant fidgeting and unpredictable activity stressful and embarrassing.

As a result of her high levels of activity and severe learning disabilities it would be hard to find full-time daycare for Maria. The social services department may eventually offer Maria two days a week in a special care unit attached to a daycentre. Finding appropriate respite care is also difficult as small care units are prepared to lock all their clients into the building and to compromise their freedom for the sake of people with Maria's problems.

Further information

Cytomegalovirus is widespread in the environment with evidence that 80 per cent of adults have suffered an infection. It is found worldwide, and is particularly prevalent in developing countries and in lower socioeconomic groups (McCance and Huether 1994). The virus may be present in semen, cervical secretions, urine, blood, saliva, breast milk and stool, and is therefore generally transmitted by close contact (including sexual transmission) or by the direct transfer of cells or bloody fluids (McCance and Huether 1994). Diagnosis can be made by cell culture. Individuals with cytomegalovirus are also at risk from super-infection from bacteria and fungi.

It is thought that 10 to 20 per cent of babies who are infected by maternal transmission become learning disabled (Clarke and McCree 1985). Indeed, the virus is now thought to be responsible for more cases of amentia than maternal rubella (Heaton Ward 1975). The clinical features of affected baby's areas described in this case study include premature birth, breathing difficulties, jaundice, hepatosplenomegaly and thrombocytopenic purpura (Heaton Ward 1975).

Main reference

Clancy, J. and McVicar, A. J. 1995 *Physiology and anatomy: a homeostatic approach*. London, Edward Arnold.

Other references

Clarke, D. and McCree, H. 1985 *Mentally handicapped people*. London, Baillière-Tindall.
Craft, M., Bicknell, J. and Hollins, S. (eds) 1986 *Mental handicap: a multidisciplinary approach*. London, Baillière Tindall.
Glanze, W. (ed.) 1986 *Mosby's medical and nursing dictionary*. St Louis, The C.V. Mosby Company.
Heaton Ward, W. A. 1975 *Mental subnormality*. London, John Wright.
McCance, K. L. and Huether, S. 1994 *Pathophysiology*, 2nd edn. St Louis, The C.V. Mosby Company.

44

The case of a girl with Prader-Willi syndrome

Sue Sides

Learning objectives

1 To understand the basis of the Prader-Willi syndrome (PWS)

2 To recognize the complexity of the management of this disorder

3 To understand the rationale for a multidisciplinary approach to care

Case presentation

Amy is a four-year-old girl with Prader-Willi syndrome. This rare and complex genetic disorder has a variety of manifestations including obesity, short stature and hypogonadism. Amy had an uneventful normal birth but was of below average weight. She developed very slowly during early infancy and her motor development was delayed. During the first two years her height and weight gain were low but thereafter her weight increased dramatically because of an excessive appetite and consequent overeating. Overeating in people with Prader-Willi syndrome is often so extreme that it is considered to be life-threatening.

Amy has a learning disability and although she was affectionate and appeared emotionally stable up until the age of three, more recently she has developed behavioural abnormalities. Her sleep disorder and occasional temper tantrums are the most problematic. Amy is frequently sleepy during the day and difficult to settle at night. Amy's care is jointly managed by her parents and a learning disability unit care team which includes a child development

clinic and a dietician. Amy attends a local playgroup for children with special needs.

Case background

Prader-Willi syndrome is caused by a rare genetic defect, located on part of chromosome 15.[1] This genotypic disorder is particularly unusual because its phenotypic expression[2] is influenced by which parent the defective gene is inherited from. (When the defect is inherited from the mother the disorder is termed Angelman syndrome.) Common manifestations of Prader-Willi syndrome are:

* Low birth weight and delayed development in early infancy, thought to be related to the poor utilization of nutrients in uterine life
* Hypogonadism: mechanism unclear
* Learning disability: particular difficulty with numbers, time and writing

All referals are from the main reference at the end of this case study.
[1] Refer to pp.605–7 for a discussion of chromosomal defects.
[2] Refer to p.595 for a definition of genotype and phenotype.

- Aggression and self-injurious behaviour. Some research studies suggest that high levels of the neurotransmitter serotonin in the central nervous system may account for these behavioural abnormalities[3]
- Sleep disorder which usually involves excessive daytime sleepiness. The cause of this is unknown but is likely to be related to some abnormality in circadian rhythmicity[4] possibly because of high levels of serotonin. Studies suggest that abnormal circadian 'sleep–wake cycles' are a common feature of this disease and may be linked to hypothalamic malfunction
- Overeating (hyperphagia) leading to early onset obesity: the cause of the craving for food is unknown but it may also be related to abnormal serotonin functioning within the hypothalamus, which is the site of the 'hunger' centre.[5]

Care

Weight control is one of the most important aspects in the care of individuals with PWS. It can have a positive effect on both the quality of life and life expectancy by the prevention of diseases associated with obesity such as diabetes, dental caries and hypertension.

Amy's care is directed primarily at controlling her weight, as well as managing her temper tantrums.

A programme of weight control involving dietary restriction will be devised for Amy. This will be calculated in relation to her stage of development and short stature. It is likely to involve a daily energy intake of less than 1000 kcal.[6] It is important that Amy is supervised at meal times and prevented from obtaining food

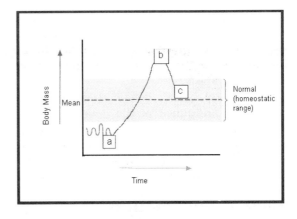

Figure 44.1 Prader-Willi syndrome. (a) Body mass below homeostatic range at birth. (b) Excessive weight caused by an over-zealous appetite. (c) Restored mass for age within normal range caused by dietary control

at other times. Dietary restriction will ensure that energy intake does not exceed energy expenditure leading to excessive weight gain.

Studies involving the assessment of total energy expenditure reveal that individuals with PWS have a significantly reduced energy expenditure as compared to normal controls. Physical exercise increases total energy expenditure and is therefore an important factor in the weight control of individuals with PWS (see Figure 44.1).

Amy will be encouraged to follow a programme of exercise as well as participating in games with other children, thereby increasing both energy expenditure and social interaction. Although beneficial, exercise programmes may be difficult to manage in children with behavioural problems and a multidisciplinary approach involving skilled play-therapists offers the best chance of success.

Family therapy will provide support and behaviour-modification techniques to enable Amy's parents to manage her temper tantrums and to allow them to continue to care for her at home. Family support is an important component of care as individuals with learning disabilities are increasingly being cared for in non-institutional environments. Family

---

[3] Refer to pp.365–9 for a general discussion of neurotransmitters and synaptic function.
[4] Refer to pp.663–4 and pp.380–1 for a discussion of circadian rhythms and sleep.
[5] Refer to p.117 and pp.121–2 for an outline of the control of food intake.
[6] Refer to pp.111–17 for a discussion of nutrient requirements in health.

therapy aims to provide a safe therapeutic environment within which relationships can be explored and effective relationships can be encouraged and developed with the help of trained therapists.

The use of behaviour-modification techniques are aimed at the development of self-control, overcoming frustration and decreasing the requirement for constant supervision.

Further information

Prader-Willi syndrome is an autosomal dominant disorder.[7] Diseases caused by autosomal dominant genes are very rare in populations. Most cases are sporadic although some affected siblings have been noted. In Amy's case there is no family history of the disease and it is therefore likely that the gene transmitted by one of the parents has undergone a mutation from a normal to a disease-producing allele. In this situation the recurrence risk (the probability that subsequent children will have the disease) is no greater than that for the general population.

Although an understanding of the genetics of Prader-Willi syndrome is not yet sufficiently advanced to enable the prevention of the disorder, genetic counselling may help Amy's parents to come to terms with her disability and may reassure them in the event of further pregnancies.

Amy's life expectancy is good although it is likely to be limited by the potential problems of obesity, rather than by the disease itself.

Some authorities advocate the use of drugs such as fluoxetine to treat behavioural problems in individuals with PWS (Helling and Warnock 1994). Fluoxetine is a selective serotonin reuptake inhibitor (SSRI) which increases the amount of serotonin available to post-synaptic neurons by inhibition of its reuptake by presynaptic neurons.

Main reference

Clancy, J. and McVicar, A. J. 1995 *Physiology and anatomy: a homeostatic approach*. London, Edward Arnold.

Other reference

Helling, J. A. and Warnock, J. K. 1994 Self-injurious behaviour and serotonin in Prader-Willi syndrome. *Psychopharmacology Bulletin* 30 (2), 245–50.

---

[7] Refer to p.598 for a discussion of dominant and recessive alleles.

The case of a man with self-injurious behaviour

45

Derek Shirtliffe

> ## Learning objectives
>
> **1** To recognize the associated causes of self-injurious behaviour (SIB)
>
> **2** To understand the context in which SIB may occur
>
> **3** To identify the purposeful nature of challenging behaviour

Case presentation

John is 45 years old. He has lived for more than 20 years in various home areas in several different hospitals for people with learning disabilities. During most of that time he has shared accommodation with other people who have self-injurious behaviours. He has limited verbal language. Assessments using adaptive behaviour scales (Nihira *et al*. 1974) indicate that he has severe learning disabilities. John has a long history of challenging behaviour which includes aggressive tendencies towards staff and other clients, destructive behaviour related to his clothing and SIB which includes head-banging.

During the day John generally avoids the company of others and sits in a chair in the home area which he regards as only his to use. In the past he has frequently hit staff who have tried to communicate with him and many now avoid him as much as possible.

Case background

Some people with learning disabilities indulge in repetitive acts directed toward themselves which result in physical harm or tissue damage. This is usually referred to as self-injurious behaviour (SIB) and commonly observed behaviours are head-banging, self-biting and self-scratching. It is generally agreed that SIB is more prevalent among people who have profound learning disabilities (Bartak and Rutter 1976).

This link with severe cerebral dysfunction has lead to some discussion of the existence of SIB because of reduced sensitivity to pain. This suggestion is supported by the observation that SIB can occur at high levels over long periods of time and result in severe injury. The idea is further supported by the observation that the individual, while oblivious to pain caused by the SIB, may be sensitive to pain associated with peripheral stimulation. This has lead to the assumption that the disorder is of a central nature associated with parts of the brain such as the thalamus, reticular formation, and areas of the limbic system.[1]

The link between SIB and reduced sensitivity to pain has not been conclusively established

All referals are from the main reference at the end of this case study.
[1] Refer to pp.615–17 for a discussion of the neurophysiology of pain.

and is further complicated by the multi-factorial nature of the perception of pain. The subjective experience of pain is influenced by the individual's unique range of anatomical, physiological, social and psychological identities.[2]

Speculation about the organic causation of SIB is further fuelled by evidence related to some rare conditions in which SIB very often occurs. An example is Lesch-Nythan disease. In this disorder it is thought that the distribution of neurotransmitters in the brain may be the cause of SIB. The area of the brain most implicated is the limbic system.

Evidence to support hypotheses which maintain that SIB arises from the biological constitution of the individual have yet to yield the exact mechanism and it is thought that organic causation is not the only determinant. Other factors which are believed to be possible contributors are:

1   Developmental hypotheses
    The appearance of SIB in normal infants has fuelled the suggestion that this is the persistence of developmentally normal behaviour (Kravitz and Boehm 1971). Some possible flaws in this argument are related to the differences in which the behaviour is manifest. For example, normal infants exhibit head-banging usually in only one situation. This rarely produces major injuries and does not persist to the same extent as similar behaviour in some disabled infants.

2   Self-stimulation hypotheses
    A possible explanation of SIB is that it serves to provide sensory stimulation which is rewarding (Bailey and Meyerson 1985). The main thrust of this argument is that SIB is essentially stereotypical in nature. Thus when confronted by an environment which is bleak in sensory experience, the SIB serves to heighten stimulation. Other theories of stereotypes suggest that the behaviour serves to reduce an individual's level of tension or arousal.

3   Learned behaviour
    Within this hypothesis the original SIB may be originated by any one of a variety of factors. The SIB is then perpetuated because it is reinforced or rewarded by some influence, usually a social response (Baumeister and Rollings 1985). Following reinforcement or reward the same behaviour is more likely to occur in the future. In the short term the reinforcement is favoured by those dealing with the behaviour because it appears to successfully resolve the SIB. In the long term, however, it can be argued that the person exhibiting SIB is being trained to exhibit the behaviour which it is hoped will diminish.

It is very difficult for observers to pinpoint the exact moment in time when an individual's SIB may originate but the factors which first produce the SIB may not be the same as those which lead to its perseverance. For example, head-banging which originated in response to pain, e.g. a headache, may persist because of the social response to the behaviour. The likelihood is that SIB is motivated and maintained by a range of permutations of these motivations which is unique to each individual.

Care

Behaviour is pushed or pulled by factors within the biological constitution or environment surrounding individuals. By their response to these influences, people are communicating a refusal to conform to the socially acceptable norms and values of the world in which they live. Thus the behaviour is purposeful and serves the individual in a number of ways:

1   To obtain something tangible that they want
2   To escape from a situation they want to avoid
3   To obtain attention from other people
4   To increase the sensory stimulation levels in the environment

[2] Refer to pp.618–19 for a discussion of the subjectivity of pain.

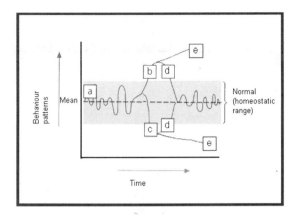

Figure 45.1 Self-injurious behaviour. (a) Behaviour within socially-acceptable norms. (b) Self-injurious behaviour as a result of biological, opperant or environmental factors leading to penalized behaviour, including possibly loss of liberty. (c) Institutionalized behaviour, leading to learned helplessness, perpetuated by the labelling and sanctioning of activities by staff. (d) Skilled nursing intervention, e.g. the positive reinforcement of appropriate behaviours to reduce inappropriate behaviour patterns. (e) Continued self-injurious behaviours caused by unsuccessful interventions.

The aims of the intervention are to increase John's independence by promoting skills which are socially acceptable to others while allowing him to retain individuality (see Figure 45.1)

An assessment of John's behaviour was carried out to establish whether it served any useful functions for him. It was discovered that the SIB was most likely to be exhibited when he wanted to obtain something, for example, to eat fruit which he seemed to enjoy and to avoid some activities which he did not seem to relish, for example, bathing. A number of strategies were employed with John to reduce the SIB:

1 Staff began to develop his communication skills primarily by beginning to teach him Makaton sign language. The patterns of staff avoidance of John were acknowledged and reversed. Thus his overall sensory stimulation was increased and the attention he received was not exclusively focused on his episodes of SIB. This activity also served to give John alternative skills in which to express himself other than by means of SIB.

2 Staff reviewed the conditions surrounding bathing which John seemed to dislike. The process of bathing was made more leisurely by allowing John to have more control over when he bathed and how long the activity took to complete. The activity was made more fun by the use of waterplay.

3 John was given greater access to food which he enjoyed, for example fruit. Previously this access had been exclusively controlled by staff.

Further information

Nurses should be constantly vigilant of the impact they have on the lives of vulnerable people. Much of nursing intervention is based on the value base of the practitioner and the individuality of clients can be suppressed via control. Nurses should admit these tendencies and implement systems which seek to diminish them such as self and citizen advocacy and referral to ethics committees which can promote an objective view of the care delivered.

Main reference

Clancy, J, and McVicar, A. J. 1995 *Physiology and anatomy: a homeostatic approach*. London, Edward Arnold.

Other references

Bailey, J. and Meyerson, L. 1985 Effect of vibratory stimulation on a retardate's self-injurious behaviour. In Murthy, G. and Wilson, B. (eds) *Self-Injurious Behaviour*. Kidderminster, BIMH Publications.

Bartak, L. and Rutter, M. 1976 Differences between mentally retarded and normally intelligent autistic children. *Journal of Autism and Child Schizophrenia*, 6, 109–20.

Baumeister, A.A. and Rollings, J.P. 1985 Self-injurious behaviour as learned behaviour. In Murthy, G. and Wilson, B. (eds) *Self-Injurious Behaviour*. Kidderminster, BIMH Publications.

Kravitz, H. and Boehm, J.J. 1971 Rhythmic habit patterns in infancy: their sequence, age of onset, and frequency. *Child Development*, 42, 399–413.

Nihira, K., Foster, R., Shellhaas, M. and Leland, H. 1974 *AAMD Adaptive Behaviour Scale, 1974 Revision*. Washington, DC, American Association on Mental Deficiency.

The case of a man with learning disabilities leaving a long-stay institution

Derek Shirtliffe

Learning objectives

1 To appreciate the necessity of circadian rhythms to wellbeing

2 To understand the impact on circadian rhythmicity of moving from a long-stay hospital setting to a community on clients who have learning disabilities

3 To identify the effects of institutionalization on clients

Case presentation

Simon is a 50-year-old man with a moderate learning disability as established by assessment using adaptive behaviour scales (Nihira *et al.* 1974). He has lived in a hospital for people with learning disabilities for the last 40 years. During his stay in the hospital Simon has lived in a variety of settings. For the past five years he has shared accommodation with 14 other men who have various learning disabilities.

In the past two weeks Simon has left hospital in order to live in an ordinary home with four other people who have learning disabilities. With the goal of Simon's complete independence in mind, staff have been working with him to determine what his wants and needs are associated with his future life. Following discussion,

Simon has revealed that he is experiencing a number of problems after his move from the hospital environment. The problems are principally connected with disturbed sleep, lack of appetite and general tiredness. These signs and symptoms are indicative of circadian rhythm desynchronization.[1]

Case background

It has been recognized for some time that many people with learning disabilities do not require care which necessitates their stay in a hospital

All referals are from the main reference at the end of this case study.
[1] Refer to pp.668–82 for a detailed discussion of circadian rhythm desynchronization.

setting because they are not subject to an on-going acute illness of any kind. A substantial number of people with learning disabilities have spent most of their lives in hospital wards.

Resettlement means for many people the giving up of routines and a way of life which has been theirs for most of their childhood and adult life. In many cases, this traditional regime of care has involved practices which have perpetuated the dependence of clients on carers and leads to the erosion of their ability to make decisions. The phenomenon of community care will have great impact on the homeostasis of the individuals involved and thus can be expected to cause circadian rhythm desynchronization (Figure 46.1).

Circadian rhythms refer to events which are repeated in the body every 24 hours. Humans are not an obviously rhythmic or cyclical species as they have no breeding season, migration or hibernation. However, some stable rhythms can be observed in humans; some of these are temperature, heart activity and metabolic rates. When these rhythms follow their expected patterns, the body is said to be in a state of internal synchronization; however, when disturbance occurs this can lead to ill-health.

This may be caused by a combination of physiological and sociological factors. Physiological factors may include damage to areas of the brain which control circadian functions, such as the hypothalamus.[2] Psychophysiological factors include changes in the daily rhythms associated with activities such as eating, sleeping, body temperature and performance. The potential for desynchronization of circadian rhythms is greater in people who are dependent on set exogenous cues to maintain their homeostasis and have greater difficulty in adaptation to new patterns of rhythm. Both these factors apply to Simon, given that he has been subjected to institutional practices. His learning disability also reduces his capacity to interpret and react to the changes in his lifestyle.

[2] Refer to pp.666–7 for an explanation of the location of circadian control.

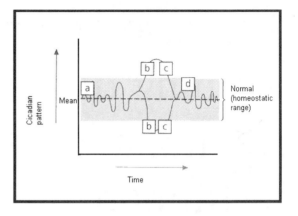

Figure 46.1 Circadian rhythm desynchronization. (a) Parameters fluctuating within homeostatic range established by circadian rhythmicity, according to exogenous cues provided by the hospital environmental cues. (b) Circadian pattern desynchronized by the loss of some of the institutionally-imposed cues and by the incorporation of new exogenous cues provided within the community. (c) Nursing intervention to re-establish synchronization, i.e. in the short-term to encourage sleep/wake and mealtime patterns as in hospital. Gradually encourage the establishment of rhythms based on new cues. (d) Parameters fluctuating within homeostatic range established by circadian rhythms synchronized to cues provided by the community environment.

Care

In order to desensitize Simon to the impending changes associated with the move from hospital to community, both the staff and Simon have agreed that the move should be gradual. With this in mind, a number of occasions where Simon has stayed in the new home have been arranged and this has continued for several months with the duration of the stay being gradually increased. Thus in the first instance Simon attended the home in order to have a meal with some of the clients who already live there. Following this Simon has stayed overnight in the home and this was further expanded to a stay lasting the whole of a weekend.

Simon has now been resident in the home for approximately two weeks. During this time he has experienced poor sleep and lack of appetite. Staff have discussed the matter with Simon and it is agreed that these problems are associated with the differences between his current lifestyle and his previous residence in the hospital environment. For example, he was previously woken in the morning by carers at an early hour because of the necessity to eat breakfast and leave the home area to attend daycare. He now has the responsibility for this activity himself and if he wishes he is free to remain in bed longer. Simon has found that he is unable to sleep beyond the time dictated by his previous routine. Equally, he ate meals according to a schedule devised for the convenience of the hospital kitchen which prepared all meals. In his new environment he is free to have meals when he wants or in cooperation with other residents in the home.

In order to compensate for the desynchronizing effect of the shift between the long-established routines and those which are new, staff have agreed with Simon to effect a number of changes to enable him to achieve homeostasis once more (Figure 46.1). It is expected that Simon's sleep pattern may adapt over a longer exposure time but short-term strategies have involved him having a short nap around lunchtime in the daycare service he attends. This has enabled him to avoid the fatigue, caused by lack of sleep, which he previously experienced. Another strategy which has been used is Simon's adoption of a programme involving cycling to improve his level of all-round fitness. The benefits of this are an increased resistance to fatigue and a general raising of Simon's feeling of wellbeing.

Further information

The main thrust of the care is aimed at transferring the locus of power away from its traditional location in the hands of carers back into the prerogative of clients. Much of the creation of power is associated with the possession of information and subsequent decision-making. Institutions have reinforced tendencies whereby clients are acted on *by* carers rather than *with* carers. Institutions as closed communities allow the skewing of client's rights with staff able to diagnose disagreement by clients as challenging behaviours to be eliminated. Practices such as block treatment, rigidity of routine and social distance serve to erode the ability of clients to become assertive and characteristics such as conformity and submission are reinforced.

Community care in a general sense removes some of the stigmatization of clients by emphasizing their rights to the same human value as other members of the community. The small scale of the home in which Simon now lives makes privacy and dignity more possible and avoids the distress and dehumanization which may accompany overcrowding. However, institutional values may prevail even in a small-scale service and it remains for the audit of the service to continue to reveal where the independence of clients can be further evolved. Where this evolution takes place it is likely that clients will experience desynchronization and need nursing intervention for this to be resolved.

Main reference

Clancy, J, and McVicar, A. J. 1995 *Physiology and anatomy: a homeostatic approach*. London, Edward Arnold.

Other reference

Nihira, K., Foster, R., Shellhaas, M. and Leland, H. 1974 *AAMD Adaptive Behaviour Scale, 1974 Revision*. Washington, DC, American Association on Mental Deficiency.

Further reading

Baxendale, S., Clancy, J and McVicar, A. J. 1997 Circadian rhythms. *British Journal of Nursing* 17 (1) 17–26.

47 The case of a baby boy born with mucopolysaccharidosis

Christine Nightingale

Learning objectives

1 To understand the genetic basis for the inheritance of mucopolysaccharidosis

2 To examine the impact of mucopolysaccharidosis on homeostasis

3 To understand the rationale for care in relation to disordered physiology

Case presentation

Bertie is an eight-year-old boy with mucopolysaccharidosis, type I Hurler's syndrome. He was born after an uneventful pregnancy to parents who were delighted at the arrival of their first son, a second child. As far as the midwife and GP were concerned Bertie had no notable or obvious abnormalities. However, after the first six months, developmental delay became noticeable. Bertie was slow to sit up and did not start rolling or moving around until he was one year old. Nevertheless he was interested and responsive and his parents were relaxed and showed no sign of anxiety. Bertie was rarely taken to health visitor clinics as his mother was busy with two children and so he was only seen for the statutory checks.

It was the health visitor who noted by formal measuring that Bertie's head was growing too large, and by observation that his facial bone structure was thickening, giving him an older worried appearance. His growth rate was also beginning to cause concern. Although his parents were short in stature Bertie's growth progress was notably slower than that recorded for his older sister.

A paediatrician diagnosed mucopolysaccharidosis. Bertie's parents were fully informed of the diagnosis and were warned to expect a marked deterioration in his condition and not to expect him to reach his teens.

At eight Bertie is totally dependent on his parents and older sister. As a two-year-old he was relatively mobile, able to crawl and bottom-shuffle his way across a room in pursuit of a desired object. He was alert to his environment and enjoyed pursuing the family cat, as many toddlers do. Now he relies for mobility on a wheelchair which is operated by his family. He shows no interest or desire to move about, and tends to stay seated on the sofa for hours on end, rocking gently. His focus appears to be on

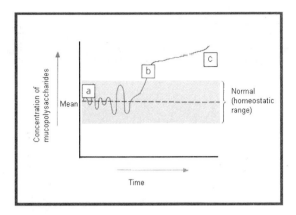

Figure 47.1 Mucopolysaccharidosis. (a) Reserve of mucopolysaccharides within cells held within normal homeostatic limits through the activities of specific enzymes. (b) Accumulation of mucopolysaccharides in an inherited disorder such as Hurler's syndrome in which certain enzymes are missing. (c) Continued deposition of mucopolysaccharides in connective tissues and endothelial cells, leading to progressive multiple organ dysfunction and skeletal abnormalities.

his knees. Unlike six years earlier, Bertie takes little to no interest in his environment.

Case background

Mucopolysaccharidosis is characterized by the abnormal storage of mucopolysaccharides (carbohydrates) in the connective tissue.[1] This causes degeneration (see Figure 47.1) and is particularly notable in the bony structures where the clinical features are severe and match the common and descriptive name for the syndrome: gargoylism. The head is large and heavy in proportion to the body, the child's face becomes coarser and frontal bossing is notable. This is accentuated by prominent supra-orbital ridges, bushy eyebrows and a depressed bridge of the nose. Ossification is delayed resulting in short limbs and stumpy hands. There is also clouding of the cornea and the individual often has a large,

All referals are from the main reference at the end of this case study.
[1] Refer to pp.70–5 for a definition of connective tissues.

fissured and protruding tongue. Urine-testing will show mucopolysaccharides.

There is a progressive deterioration in mental and physical function. Disorders usually involve the liver, spleen, heart and blood vessels (McCance and Heuther 1994). The child is not expected to live beyond ten years.

Providers of care need to recognize the progressive nature of this syndrome. While striving for continued learning and independence for the service-users, such as Bertie, they must work with the limitations imposed by the syndrome. An understanding of the organs and tissues primarily affected by deterioration will help the carer assess and recognize the progressive problems and alert medical personnel for appropriate pain relief and treatment, including regular physiotherapy to relieve the pain and discomfort of joint contractures.

Support for the family is vital and they should be given opportunities to discuss the implications of the disorder. Carers need to prepare the family, including other siblings, for an early death. Sufficient information should be made available to enable the parents to make decisions about genetic counselling[2] and the options available to them if they wish to have another child.

Further information

Mucopolysaccharidosis manifests itself in at least seven varieties. Three are recognized as most commonly associated with learning disabilities: Hurler's (type IH), Hunter's (type II) and Sanfilipo (types IIIA and IIIB) syndromes.

All except Hunter's (type II) are inherited autosomal recessive disorders;[3] Hunter's is X linked[4] and therefore has implications for female relatives and their male children.

[2] Refer to pp.601–7 for a discussion of gene inheritance.
[3] Refer to pp.601–3 for a discussion of autosomal recessive inheritance.
[4] Refer top.604–5 for a discussion of sex-linked inheritance.

Main reference

Clancy, J. and McVicar, A. J. 1995 *Physiology and anatomy: a homeostatic approach*. London, Edward Arnold.

Other reference

McCance, K. L. and Huether, S. 1994 *Pathophysiology*, 2nd edn, St Louis, Mosby.

The case of a girl born with cretinism

Gabriel Ip

Learning objectives

1 To understand the normal metabolic functions of thyroxin

2 To appreciate the symptomatology of cretinism

3 To understand the role of hormone-replacement therapy

Case presentation

Margaret Bracknell was born ten years ago as the only child to the Bracknell family. Her mother did not attend ante-natal clinic during her late pregnancy. Margaret was born with a low birth weight and has had a spate of prolonged jaundice. She has coarse facial features: broad flat nose, thick lips and a large protruding tongue. Umbilical hernia was also reported but it was corrected by surgery.

Margaret is a quiet little girl who is described as 'slow' in her actions and reactions and she shows little interest in her surroundings. Margaret is also prone to constipation and did not feed well as an infant.

In her early schooling Margaret is often a target for 'bullying'. Concerns from her parents brought forth assessments by educational psychologists. Findings of an IQ of 74 (with the WISC-R test) and an IQ of 56 (with the Stanford-Binet) were presented. This prompted her parents to press for further investigations on patho-physiological lines.

Margaret has a 'puffy' look and generally her skin is flabby: it feels cold and dry to touch. There is evidence of delayed skeletal development and

muscular weakness. A basal metabolism test shows a 35 per cent reduction from the reference group: in fact, a consistent low body temperature of <35°C was found.

From the case history it was identified that Margaret has cretinism (also known as primary congenital hypothyroidism) (PCH) possibly caused by a recessive autosomal inheritance from her parents.[1]

Case background

The thyroid gland secretes tri-iodothyronine and tetra-iodothyronine (abbreviated T_3 and T_4 respectively). T_4 is the main secretion and is commonly called thyroxine. The secretion of the hormone thyroxine is maintained by two negative feedback loops: the long and/or short loops.[2]

All referals are from the main reference at the end of this case study.

[1] Refer to pp.601–3 for a discussion of autosomal recessive inheritance.

[2] Refer to Figure 15.11c and p.442 to understand the relationship between thyroxine and TSH negative-feedback actions.

These are modified amines, derived from the amino acid tyrosine. Their names reflect the number of iodine atoms that are incorporated into each hormone molecule.[3] Thus, iodine must be included in our diet. The iodide ion is actively taken up by the gland and initially incorporated into tyrosine residues attached to a protein called thyroglobulin[3] The uptake of administered radioisotopic iodine is a clinically useful means of monitoring thyroid function.

The main function of thyroxine is to determine the basal metabolic rate. The symptoms suggest an overall retardation of both physical and mental functions but these are only apparent as child development progresses and functions mature, particularly in early childhood. This condition is not easily recognizable at birth, even by the most experienced paediatrician.

Thyroxine has a profound effect on the metabolic processes of the human body: it acts as a catalyst for the oxidative reactions of body cells by converting food into new body cells and into energy. It also regulates the rate of oxygen consumption and hence the metabolic rate (conventionally measured by the basal metabolic rate: BMR).

Care

In suspected cases of cretinism, clinical investigations include serum PBI (protein bound iodine), total serum T_4, plasma TSH, radioactive iodine studies, skeletal radiological studies and neonatal screening.

On confirmation that the secretion of the hormone thyroxine is deficient, homeostasis is restored by artificially replacing the hormone thyroxine. This therapeutic replacement keeps the hormone level above the threshold (see Figure 48.1) and therefore no symptoms

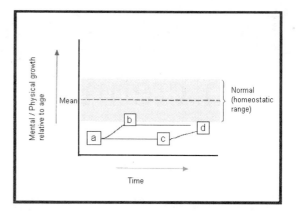

Figure 48.1 Cretinism. (a) Inadequate mental/physical growth at birth associated with low levels of serum thyroxine as occurs in congenital hypothyroidism (cretinism). (b) Intervention following early diagnosis results in improved mental/physical growth, although not equal to age-related norms. (c) Intervention following later diagnosis resulting in a slight improvement in physical and mental growth, as in (d).

appear. For many years, the standard preparation has been dried thyroid gland extract. Nowadays, there is a pure hormone available as L-thyroxine sodium (eltroxin) 100μg tablets. The usual replacement dosage is between 300 and 400 μg daily. This preparation is more potent than the dried extract, as one tablet equals 60 mg of dried extract.

The other care implications are that carers should be sensitive to Margaret's needs in these areas:

- promote fluid, fruits and other roughage in order to minimize the risk of constipation
- supervise closely to prevent accident: people with PCH are known to 'lean against hot radiators to keep warm'
- observe and minimize the risk of hypothermia

As with the majority of inherited disorders there is no active treatment programme to reverse this condition. The emphasis is put on:

- an early diagnosis

---

[3]Refer to Figure 15.2 for an examination of the structure of tyrosine and its derivative amine hormones, and to Figures 15.11a and 15.11b for the thyroid synthesis of T_3 and T_4.

- the replacement of thyroxin
- maintaining an acceptable serum thyroxin
- the continuous monitoring of serum T_4 and PBI
- a collaborative multidisciplinary approach
- genetic counselling

Further information

In hypothyroidism if the deficiency of thyroxine secretion is not remedied and is allowed to fall below the threshold, symptoms appear in increasing frequency and continue to severity.

The overall incidence is between 1 in 4000 and 1 in 7000. Arguably it is six times as much as the incidence of phenylketonuria (Edwards et al. 1995). It could also be an inborn error of metabolism of the thyroid hormones: iodine deficiency, pituitary or hypothalamic disorders and ingestion of goitrogens in utero. During pregnancy, maternal hormones cannot compensate for fetal-thyroid deficiency. As a result, the entire metabolic process of the developing fetus is slowed down. Stillbirth is commonplace. If the child does develop to term there are great risks of retardation, both mentally and physically (McFaul et al. 1978).

Early diagnosis is paramount here (Miculan, Turner and Paes 1993) as delay in starting replacement therapy will inevitably lead to permanent, irreversible and impaired cerebral functions. The longer the delay, the more severe the damage.

Maintenance of replacement therapy through L-thyroxine: this may be administered orally throughout life. Some authors suggests that once puberty is reached the dose can be gradually reduced, as cerebral function is established (Clarke, Clarke and Berg 1985).

Main reference

Clancy, J, and McVicar, A. J. 1995 *Physiology and anatomy: a homeostatic approach.* London, Edward Arnold.

Other references

Clarke, A., Clarke, A. and Berg, J. (eds) 1985 *Mental deficiency – the changing outlook.* London, Methuen, 199–201.

Edwards, C., Bouchier, I., Hasle T. C. and Chilvers, E. (eds) 1995 *Davidson's principles and practice of medicine.* London, Churchill Livingstone, 692–3.

McFaul, R., Dorners, S., Brett, E. and Grant, D. 1978 Neurological abnormalities in patient treated for hypothyroidism from early life. *Archives Diseases of Childhood,* 53, 611–18. In Clarke, A., Clarke, A. and Berg, J. (eds) 1985 *Mental deficiency – the changing outlook,* Chapter 5. London, Methuen.

Miculan, J., Turner, S. and Paes, B. 1993 Congenital hypothyroidism: diagnosis and treatment. *Neo-Natal Network,* September, 12(6), 25–34.

49 The case of a man with phenylketonuria

Gabriel Ip

> **Learning objectives**
>
> **1** To understand the basis of phenylketonuria (PKU)
>
> **2** To recognize the rationale for therapeutic intervention in PKU
>
> **3** To increase an awareness of the role of genes in health
>
> **4** To increase an awareness of the principles of the inheritance of genetic disorders

Case presentation

Jim is 36 years old. His birth had been uneventful and normal. Unlike his parents, Jim has light-blonde hair, blue eyes and a pale complexion. As a child, he was described as being irritable, immature and overdependent. He was 'hyperactive', and rocking back and forth, waving his arms and grinding his teeth were also not uncommon. His skin and urine were often described by his parents as having a 'musty' or 'mousy' odour.

At 16, Jim was diagnosed as having severe learning disability. At times he suffered from epileptic seizures, dystonia and dysphagia. His behaviour had become too trying for his parents and consequently he was received into care at a local residential home. Neither parent exhibits the disorder, nor does his sister.

Jim was subsequently diagnosed as having phenylketonuria.

Case background

Phenylketonuria is a condition which results from an individual's inability to convert the amino acid phenylalanine to tyrosine as a consequence of a deficiency of the enzyme phenylalanine hydroxylase.[1] Phenylalanine concentration in the blood then becomes elevated (Figure 49.1). The signs and symptoms exhibited by Jim arise because:

- excess phenylalanine interferes with the entry of other amino acids into the brain across the 'blood-brain barrier'.[2] A full range of amino acids is required for normal protein synthesis, and so amino acid deficiency interferes with brain development in PKU.

All referals are from the main reference at the end of this case study.
[1] Refer to pp.34–8 and pp.49–51 for a discussion of the role of enzymes in health.
[2] Refer to pp.357–60 for a discussion of cerebrospinal fluid and blood-brain barrier.

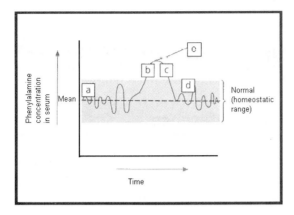

Figure 49.1 Phenylketonuria. (a) Serum phenylalanine concentration within its homeostatic range, i.e. dietary intake and utilization of the amino acid are in balance. (b) Rising phenylalanine concentration is a result of inadequate utilization. (c) Intervention to reduce phenylalanine concentration and restore homeostasis by reducing dietary intake as in (d) (i.e. intake now in balance with reduced utilization). (e) Persistent elevated phenylalanine concentration in the serum. This will result in the signs and symptoms of phenylketonuria caused by the inadequate dietary control. The neurological damage which results is irreversible

- tyrosine deficiency in the brain prevents the normal production of adrenergic neurotransmitters, such as noradrenaline, and so affects synapse development.[3]
- excess phenylalanine promotes the production of phenyl-pyruvic acid (also known as phenylpyruvate) and this 'waste' substance is responsible for the 'musty or mousy' odour of sweat and urine.
- tyrosine is a precursor for the skin pigment melanin. Its deficiency, therefore, produces fair hair, blue eyes and a pale complexion.

The enzyme deficiency responsible for phenylketonuria arises because of a genetic mutation (the condition is inherited) and the disorder is currently irreversible.

[3] Refer to pp.365–9 for a discussion of synapses and neurotransmitters.

Care

The prospects for Jim's return to health are poor. Care is therefore directed at maintaining and promoting as much self-care as possible.

The condition can, however, be controlled if it is diagnosed early, and babies born in Britain today are routinely tested for phenylketonuria. A simple blood test (the Guthrie test) soon after birth will detect if phenylalanine concentration is elevated. If it is elevated the child will then be placed on a low-phenylalanine, or phenylalanine-free diet. Tyrosine supplements may be required if this is extremely deficient. The education of the parents, and the child when they are old enough to understand, to the necessity of this controlled diet is essential (Figure 49.1).

Care is therefore directed at normalizing the phenylalanine concentration in body fluids. Since the concentration is determined by the balance between phenylalanine uptake and its utilization (it is an 'essential' amino acid, so it cannot be synthesized or stored by the liver[4]). Phenylketonuria can be controlled by restoring this homeostatic balance. The utilization of phenylalanine cannot be promoted (nor is correction of the underlying gene defect possible), so therapy is directed at reducing the dietary intake of the amino acid (and increasing the intake of tyrosine if necessary).

Other points to note:

- the irreversibility of the damage caused by the condition means that control must begin at an early age before brain development has progressed to an advanced stage.
- phenylalanine is abundant in natural food proteins. Dietary control of phenylketonuria is therefore highly restrictive.

[4] Refer to p.153 for notes on the processes of transamination/deamination.

Further information

Phenylketonuria is an inherited condition. Its absence in both of Jim's parents indicates that it is autosomal recessive in nature, so Jim must be homozygous for the condition while both parents are heterozygous.[5] The statistical likelihood of the children of heterozygous parents being homozygous for a recessive gene is one in four or 25 per cent.[6] Jim's unaffected sister may be heterozygous like the parents (i.e. a carrier of the phenylketonuria gene) or homozygous for the normal phenotype.[7] The worldwide incidence of phenylketonuria is 1 in 10 000 live births. Geographic and/or cultural factors will influence these figures, however.

Main reference

Clancy, J. and McVicar, A. J. 1995 *Physiology and anatomy: a homeostatic approach*. London, Edward Arnold.

[5] Refer to pp.598–9 for a discussion of inheritance.
[6] Refer to pp.601–2 for a discussion of autosomal recessive inheritance.

[7] Refer to p.595 for notes on 'phenotype'.

Index